THE SUN

Popular Science

THE SUN

PARUL R. SHETH

Illustrations
BELA MODI

nbt.india
एकः सूते सकलम्
NATIONAL BOOK TRUST, INDIA

ISBN 978-81-237-4424-7

First Edition 2005
Reprints 2005, 2008, 2009, 2011, 2012, 2013, 2014, 2017, 2021 (twice), 2023 (*Saka* 1945)

₹ 170.00
Published by the Director, National Book Trust, India
Nehru Bhawan, 5 Institutional Area, Phase-II
Vasant Kunj, New Delhi-110 070
www.nbtindia.gov.in

Contents

Acknowledgements

First and foremost, I am grateful to the National Book Trust, India for giving me the opportunity to write this book. I would like to extend my thanks to Ms Manju Gupta and to Ms Kanchan Wanchoo Sharma who imparted an untiring enthusiasm that made me complete this book. Their valuable suggestions and excellent editing skills have helped in the making of this book. Indeed, bringing out any publication is not a one-man show. It is a joint effort of many individuals having expertise in various fields, and hence I am thankful to all the staff of NBT for a splendid job.

I am indebted to Dr. S.M. Chitre, former senior professor with the Theoretical Astrophysics Group, Tata Institute of Fundamental Research (TIFR), Mumbai, for so readily sparing his valuable time and blessing this work with an apt foreword.

I am especially thankful to Mr. Suhas Naik-Satam, Coordinator, Nehru Planetarium, Mumbai, for his invaluable suggestions and comments, and for his ever-helping hand that made this endeavour successful. It is difficult to name all the well-wishers and friends who spared their precious time and helped me add a new dimension to this venture. I am highly obliged for their invaluable suggestions and help.

My special thanks go to Dr. Bela Modi, a dentist by profession, and a wonderful artist who has painstakingly prepared the illustrations in record time.

The acknowledgements cannot be complete without thanking each one of my family members who stood by me, and without whose cooperation it would have been impossible to go through the rigours of writing this book. I would specially like to dedicate my work to my parents, Ghanshyam and Jayavanti, my sisters, Bela and Madhavi, Rajen, my loving husband and Miten, my son, the most loved man in my life.

PARUL SHETH

Foreword

Since ancient times, humankind has been conscious that the life on Earth is sustained by the Sun. It is, therefore, not surprising that through the ages many cultures have held the Sun in great reverence and some societies have even worshipped it as God!

In this fascinating book on the Sun, Dr. Parul Sheth has given a very illuminating and lucid description of the central role played by our star in influencing the course of events on this planet, affecting not only the daily human life but also controlling the plant and animal lives around us. After a scrupulous description of the Sun and its family of planets and minor bodies, there follows an up-to-date account of the physics of the solar interior and exterior, particularly highlighting various transient phenomena, recorded by ground-based and satellite-borne instruments, occurring on the Sun's surface and in its overlaying atmosphere. A noteworthy section in the book addresses the important question of how the Sun shines using the source of energy generated by thermonuclear reactions taking place in its central regions and traces its evolutionary life-history until it has exhausted this prodigious energy source to end up finally as a white dwarf.

Man has always been curious about various phenomena occurring in nature like changing colours of the sky at sunrise and sunset, dazzling display of aurorae in the polar latitudes, formation of colourful rainbows and spectacular view of the solar chromosphere and corona during total eclipses of the Sun. A convincing case is made as to how these natural events can be eventually attributed to our Sun. The author has also briefly described the role of photosynthesis, with respect to the terrestrial environment, when food is produced by plants using energy obtained from the Sun. The use of sunlight for navigation by migratory birds is also suitably described.

The author has provided a fairly exhaustive list of energy sources that could be tapped for commercial purposes using winds, tides, ocean currents, passive and active solar heating, photovoltaic cells and bio-energy to supplement the hydroelectric, geothermal and nuclear energy, but finally comes to the conclusion that the ultimate source for providing the ever-increasing human energy requirement has to depend on the direct conversion of solar energy which will, hopefully, become an economically viable proposition in not too distant future. The last few sections highlight the Sun's contribution to the promotion of health with a cautionary note on the ill-effects of the excessive exposure to sunlight.

It has been an enjoyable and instructive experience for me to go through this well-documented and readable account of our star to which plant Earth is closely linked. And this same star has made possible the very existence of our Earth and its ability to nurture sustainable and flourishing variety of life-systems in its environment. Clearly it is vital for us to identify the processes that influence the Sun-Earth relationship, and which help us to understand how variations in the solar radiation output affect the weather, global climate and ozone content in our atmosphere, and how the energetic events occurring on the Sun disrupt communications and power supply on Earth. After all we have to learn to live with our star, and it is, therefore, essential for us to identify possible solar drivers that are responsible for explosive release of energy, causing intense disturbances that approach Earth with great speed and eventually affect societal lives. I hope this book will stimulate the readers to undertake serious pursuit of the Sun and encourage them to wonder about some of the outstanding problems in Solar Physics.

DR. S.M. CHITRE
Former Science Professor
Theoretical Astrophysics Group
Tata Institute of Fundamental Research, Mumbai

solar flares that are responsible for explosive release of energy, causing intense disturbances that approach Earth with great speed and even cripple artificial satellites. I hope this book will stimulate the readers to undertake serious pursuit of the Sun and encourage them to wonder about some of the outstanding problems in Solar Physics.

DR S M CHITRE
[illegible] Professor
Theoretical Astrophysics Group
Tata Institute of Fundamental Research, Mumbai

Preface

For centuries, human beings have attempted to portray the Sun according to their own worldviews. The Sun has been thought of as, a God, a demon, and a mischievous spirit according to the different folklores from around the world. It is the Sun that makes our Earth a warm and a comfortable place to live on. It provides us the heat and creates daylight by emitting electromagnetic radiation. Plants use the light from the Sun to grow and to make food via photosynthesis, which in turn absorb carbon dioxide and give out oxygen. Light and heat from the Sun are the prime energy ingredients necessary for growth, movement and reproduction in nearly most life forms on the Earth.

Think about what would our world be like, without the Sun. The Sun influences our entire lives and also our environment. Besides the many aspects of purely scientific research, there are numerous natural and environmental influences caused by the Sun. We have learnt to make use of the Sun's energy in various ways. Right from lighting and heating our homes, and using solar power for cooking, to using solar energy for shortwave radio communications, satellites, navigation, electrical power transmission, and so on—the Sun plays an important part in our development as a modern social society.

One of the major areas where the Sun's contribution is often discounted is in its promotion of health. Without the Sun, life would not exist. Sun is our friend, but too much exposure to Sun—extreme heat, intense light and solar radiation—can prove hazardous to our health. This book is written with the intention to guide you through the varied aspects of the Sun. So hop on aboard and enjoy the ride to 'The Sun'.

1

OUR STAR IN THE COSMOS

Among the billions of stars in the Milky Way galaxy, there is one that we can call our star, the Sun. Our Sun appears larger than the other stars that we see in the sky at night. This is because it is the nearest star from the Earth. Although the Sun seems like any other ordinary star, it is one of the most complex things in the Universe. It gives light and heat to many planets. It provides many elements essential to life on Earth. This hot ball of molten plasma heats our planet, creates our weather, and plays a significant role in improving our health. This fiery mass is also the driving force behind harsh weather patterns, life threatening illnesses, and even disruptions in our lives.

In spite of causing problems at times, the Sun is mankind's best friend. It is a life-giver. It benefits all living things. It provides fresh air, initiates flora and fauna, improves human health, makes plants grow, creates rainbows, creates the seasons, provides solar energy, enhances communications, keeps the Earth warm, helps us tell time, cleanses water and shorelines. The list is endless.

'Old Sol' as scientists refer to the Sun, plays a powerful role in our everyday lives. From the beginning of time it has influenced, moulded, and helped shape many cultures.

History and Mythology

The sky in ancient times seemed to be mystical—the home of gods, demons and spirits. Great fear encroached upon the minds of people when events such as eclipse, comets or meteor showers occurred in the sky. The ancient Egyptians about 4,000 years ago looked on the Sun as a creator-God (Fig. 1). Their legends tell us about a vast world-ocean called *Nun*. *Atum*—the Sun god created himself out of the waters. *Atum* made a mound of Earth to stand and this became the land. Being the creator, he then created the lesser gods, one to rule over moisture and the other to rule over the atmosphere.

Ancient people thought of the Sun as a perfect sphere of celestial fire created by gods. The Egyptians from Africa, the Sumerians in Asia, the Greeks in Europe, the Aztec and Maya Indians in North America and the Incas

Fig. 1.1: Egyptian Sun worship

in South America; all worshipped the Sun as a god. The Sun is personified in many mythologies: the Greeks called it *Helios* and the Romans called it *Sol*. An old legend tells us that a sacred boat carries the Egyptian Sun god *Ra* or *Re*. The boat rises over the eastern horizon and sails across the heavenly ocean. At night the boat sinks below the western horizon and continues its journey through the underworld until it rises in the east and begins a new day. The Greeks believed that the Sun god *Helios* drove a chariot through the sky. The Eskimos too, thought the Sun took a boat ride at night beyond the northern horizon and was responsible for the *aurora borealis*, or northern lights. The Maoris from New Zealand believed that one of their heroes had fought the Sun and crippled it, so that it limped across the sky.

In archaic times, people perceived the Sun to be a passionate aspect of the Great Mother. She was the Heaven-illuminating Goddess, *Ameratsu Omikame* in Japan, and the Queen of Heaven and Earth, *Arinna*, in Mesopotamia. She was *Yhi*, Sun Woman, to the *Arunta* of Australia and *Antolia*, in Siberia and Native America. Tribal northern Europe knew her, too. The Germans called her *Sunna*, as did the Norwegians. In Scandinavia, she was *Glory-of-Elves*, or *Sol*. She was *Sol*, as well, to the Celts who also called her *Sul* or *Sulis*. Her celebrations took place on hilltops overlooking springs. A major ceremonial site was Silbury Hill (Sulisbury Hill) adjacent to the springs at Bath, once called Aquae Sulis. The Romans dedicated altars at *Bath*, sacred to *Sul Minerva*.

The *Mahanirvana-tantra* describes the Sun as a golden garment of light which graces the great goddess. The great Mother in ancient India was *Aditi*, the mother of the twelve spirits of the zodiac, the *Adityas*. Aditi is the mother of Sun, also known as *Aditya*. Hindus also worship *Randel ma*, wife of the Sun god, besides the Sun god himself. The Sun God *Surya,* is the eye of *Varuna,* the God of the Heavens. In Greece, the Sun, *Helios,* is the eye of *Zeus;* in Egypt, the eye of *Ra;* in northern Europe, the eye of *Odin;* in Oceania, the eye of *Atea;* and in Islam, the eye of *Allah.*

Hindu mythology is replete with references to the Sun god or Lord *Suryanarayana* (Fig. 1.2). The Rig Veda is studded with prayers to many gods. The maximum number of hymns in Rigveda is devoted to *Agni,* the god of fire. The Supreme god or the creator manifests himself as fire—*Agni* on this earth (*Prithvi*), as lightning or air (*Indra* or *Vayu*), in the sky (*Antariksha*) and as the Sun (*Surya*) in the heavens (*Devloka*). The Sun is sometimes described as the discus carried by Lord Vishnu, the Supreme deity of the Vaishnava sect. Hindus worship the Sun and they even observe a fast on Sundays—the day of the Sun god. The Konark Sun temple in Orissa in India is a place for all Sun worshippers. Yoga too teaches *Surya Namaskara* to be practiced every morning.

Fig. 1.2: Lord Suryanarayana

In the earlier days, people built monuments to worship the Sun. Some of these served to tell time. The Sun was used in calendar making because its presence marked the difference between day and night. Equinoxes, solstices and eclipses could be determined by looking at its position. The Egyptians made carvings of the Sun and Sun worshippers. The Incas used granite monuments as Sundials to show the direction of the Sun's shadow (Fig. 1.3). The direction changed as the Sun moved. Many ancient people built complex structures to learn about the Sun's motion from north to south and back again as the seasons changed. At Stonehenge in England, the Sun rose over a stone marker. The structure tracked the motions of the moon and the Sun. Inside the circle of stone blocks, people probably held ceremonies to salute the Sunrise and to pray for a season of good crops.

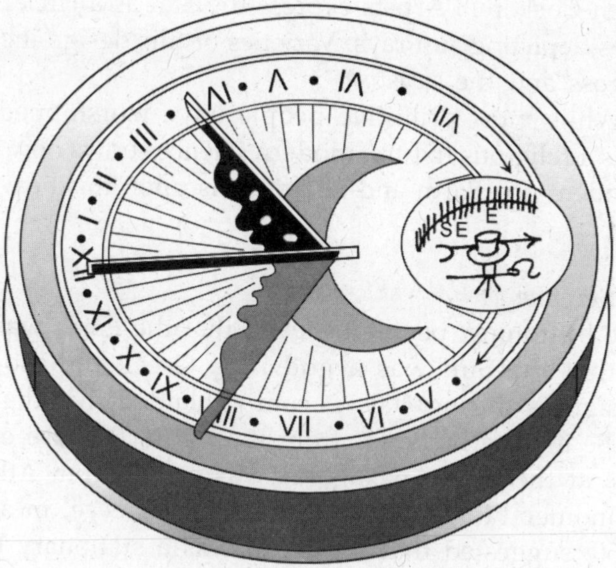

Fig. 1.3: The Sundial

The Sun's influence has not only led to many scientific discoveries, it also has affected the Arts, including music, painting, and dance. One of its greatest influences, however, is in religious culture. Many of the ancient civilizations worshipped the Sun god in some form or the other. *Apollo*—the Sun god is said to be the god of musicians and poets. He carries a lyre in one hand and the egg of creation in the other. Sun is described, as a sea of flames. This sea with its vibrant colours is a perennial favourite of artists, authors and poets. The Sun's beauty and warmth have been captured on canvas. The Dutch painter Vincent van Gogh created landscapes bathed in bright sunshine that expressed his joy. The American poet Emily Dickinson wrote a poem called *The Sun* in which she described the rising and the setting of the Sun. The Russian composer Nikolai Rimsky-Korsakov inserted a beautiful song *Hymn to the Sun* in his famous opera 'The Golden Cockerel'. Artists express the Sun as a circle with spokes depicting sun rays. Varieties of this design include the cross and the *swastika*.

While most primitive people have worshipped the Sun as a religious fixture, modern science shows how very dependent our Earth and all its life is upon this heavenly object.

Sun Science

In the beginning, people believed that the Earth was flat and that the Sun was a god. The Greek philosopher Anaxagoras during 400 BC, comprehended that the Sun may be a large body, far from the Earth. Because of his ideas, at the time Anaxagoras was exiled from Athens. Astronomer Ptolemy of Alexandria in Greece, in about AD 150, suggested that the Earth was a stationary body and all the planets, stars, the moon and the Sun, all revolved around the Earth.

Until the Middle Ages it was assumed that the Sun orbited the Earth. It was only in 1543 that the Polish astronomer Nicolaus Copernicus revealed that the Sun was at the centre of the solar system and the Earth and the other planets revolved around the Sun. However, his view of the Solar System wasn't accepted for many years until Isaac Newton formulated his 'laws of motion'. Astronomers were then able to prove scientifically that Sun was actually a star.

The quantitative study of the Sun dates from the discovery of sunspots but the study of its physical properties was initiated much later. The English mathematician and scientist Sir Issac Newton observed the Sun's spectrum in the year 1666. The German physicist Joseph von Fraunhofer in 1814 studied the properties of Sun using a spectroscope laying the foundation of a detailed theoretical explanation of the solar atmosphere.

In 1859, the German physicist Gustav Kirchhoff first showed that the lack of radiation by atoms at certain wavelengths in the Fraunhofer spectrum of the Sun was due to absorption of radiation of atoms of some of the same elements present on the Earth. This demonstrated the possibility of deriving information about celestial objects by studying the light that the objects emitted. Thus began the field of astrophysics.

In 1904, the American George Ellery Hale set up the Mount Wilson Observatory near Pasadena, California, USA where scientists could learn about the Sun and other stars. He was the one who coined the word 'astrophysics' meaning study of astronomy using methods of physics. The American astronomer Horace W. Babcock in 1948 invented the magnetograph that measures the magnetic-field strength over the solar surface. The ability and perseverence of the scientists to make newer and improved observations has led to the progress in understanding the Sun. Today, the Sun is viewed as a dynamic, evolving body and its properties and variations are being understood scientifically.

Astronomers use solar telescopes, solar spectrographs and the coronograph to study the Sun. During the 1960s, astronomers used space probes (satellites sent into outer space) to study the solar cosmic rays. Scientists believe that much of the motion in the Sun's atmosphere occurs in the form of waves. The study of waves, known as helioseismology reveals variations of temperature, density and chemical composition in the Sun's interior.

Today the study of Sun's atmosphere is possible because of newer instruments and adaptations of old ones. Earlier, the spectroscope was used to study the Sun's atmosphere but only at the time of a total eclipse of the Sun. By focusing a spectroscope on the small rim of the Sun, which remains in view just before the moment of totality, it is possible to obtain the spectrum of the lower part of the atmosphere. This spectrum is known as a 'flash spectrum', since unlike the normal continuance spectrum of the Sun, it consists of bright lines.

George Ellery Hale invented the spectroheliograph, which is an extension of a spectroscope, although a physicist named Deslandres discovered the principle independently in France at about the same time. Hale further refined the spectroheliograph into a new instrument called the spectrohelioscope. This instrument, instead of taking a photograph as the spectroheliograph does, permits one to view the solar disk and surrounding atmosphere visually. The observer could now see instantaneous and amazing changes in the photosphere. An ordinary photograph does not reveal the detailed structure of the Sun's upper atmosphere. This is due to the intense glare from the photosphere, which does not allow observation of fainter details or the higher gases of the chromosphere.

Several spacecrafts have been launched to conduct experiments and learn more about the Sun. However, none have gotten any closer to its surface than approximately two-thirds of the distance from Earth to the Sun.

The Birth of Sun

More than 100 billion stars make up the Milky Way. Throughout the Milky Way, and in the space between the galaxies, there are huge clouds of gas and dust. A new star forms when gas and dust together begin to contract under the force of gravity producing heat. As the gas and dust mask shrink, the temperature at the centre of the mass increases. The temperature at the centre becomes so high that the thermonuclear reactions begin to occur producing energy. This energy causes the mass of gas and dust to shine as a star.

Astronomers believe that the Sun was formed about five billion years ago from a rotating mass of gas and dust. Enormous clouds of dust and gas, stretching over two light years, collapsed under the force of gravity. As this material was compressed it gradually warmed up to temperatures of 15 million degrees. At this point nuclear fusion began in its core, and energy in the form of heat and light travelled out from the centre. As long as there is enough material within the star, the nuclear fusion process can continue with hydrogen being fused to form helium. The released energy causes the star—the Sun—to shine.

The Sun is an average-sized, middle-aged star but it dwarfs our planet. About 109 Earths could fit side by side across the diameter of the Sun (Fig. 1.4). Best estimates tell us that the Sun was born 5 billion years ago, and it has completed nearly 22 trips around the centre of the Milky Way at a speed close to 240 km per second. At 4.5 billion years our Sun is still young. It is estimated that there is enough fuel in the interior of the Sun to keep that lamp burning for about another five billion years.

When you look up in the sky at the Sun and the moon, you notice a strange coincidence, both look the same size in the sky. But in fact, they are not really the

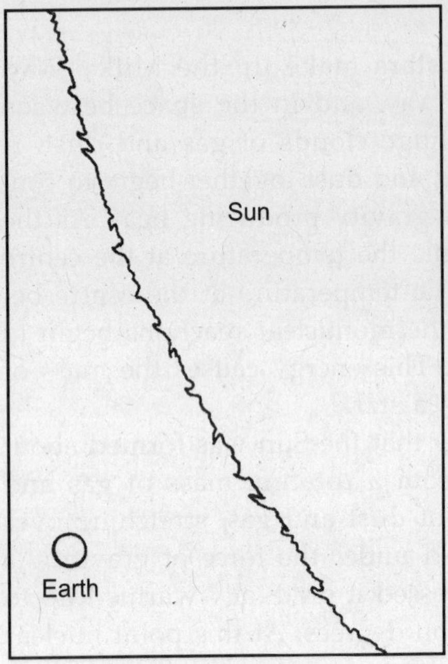

Fig. 1.4: The Sun size

same size. The Sun's diameter is actually 400 times the moon's diameter. And the Sun is 400 times further away from the Earth, reducing its apparent size to the same as the moon's.

In the descriptive terms of astronomers today, the Sun is a yellow, main sequence star that belongs to the spectral class G2, and has an absolute magnitude of +5. It is composed of 74 per cent hydrogen, 25 per cent helium, and 1 per cent other elements, more than sixty of which are naturally occurring elements on Earth. The Sun contains more than 99 per cent of all the mass of the Solar System, and it has 330,000 times the mass of the Earth.

Of all the stars in all the galaxies, it is the closest one around which our world revolves. It is quite literally, "The light of our lives", as our existence depends upon its energy. We are fortunate that the Sun is exactly the way it is. If it were different in almost any way, life would almost certainly never have developed on Earth.

The Sun is about 150 million kilometres from the Earth. Its diameter is around 1.4 million kilometres. The Sun spins around once every 27.4 days. The temperature at the core of the Sun is approximately 15 million degrees Celsius.

The coolest part of the Sun is nearly 6,000 degrees Celsius. The Sun loses approximately 4 million tons of 'matter' every second—this is the amount of hydrogen gas that the Sun turns into energy.

There are sometimes 'Mock Suns' (parhelia) which are called 'Sundogs' because they follow the Sun around. These Sundogs are usually seen as bright spots on opposite sides of the winter Sun.

The Sun's Family

The Earth is not the only large body to circle the Sun. There are eight other planets and a large number of smaller objects orbiting the Sun. Each planet circles the Sun at a different distance and takes a different time to orbit the Sun. These planets form the solar system or the main part of the Sun's family (Fig. 1.5). Several planets are themselves the centre of miniature systems with many satellites, or moons circling around them. A group of large rocky bodies called asteroids circle in a broad band, or belt, between the orbits of Mars and Jupiter. The Sun's family also includes comets and meteors, or shooting stars.

Planet	*Distance* (000 km)	*Radius* (km)	*Mass* (kg)	*Discoverer*	*Year*	*Orbits*
Mercury	57,910	2439	3.30e23			88 days
Venus	108,200	6052	4.87e24			225 days
Earth	149,600	6378	5.98e24			365 days
Mars	227,940	3397	6.42e23			687 days
Jupiter	778,330	71492	1.90e27			12 yrs
Saturn	1,426,940	60268	5.69e26			30 yrs
Uranus	2,870,990	25559	8.69e25	Herschel	1781	84 yrs
Neptune	4,497,070	24764	1.02e26	Galle	1846	165 yrs
Pluto	5,913,520	1160	1.31e22	Tombaugh	1930	248 yrs

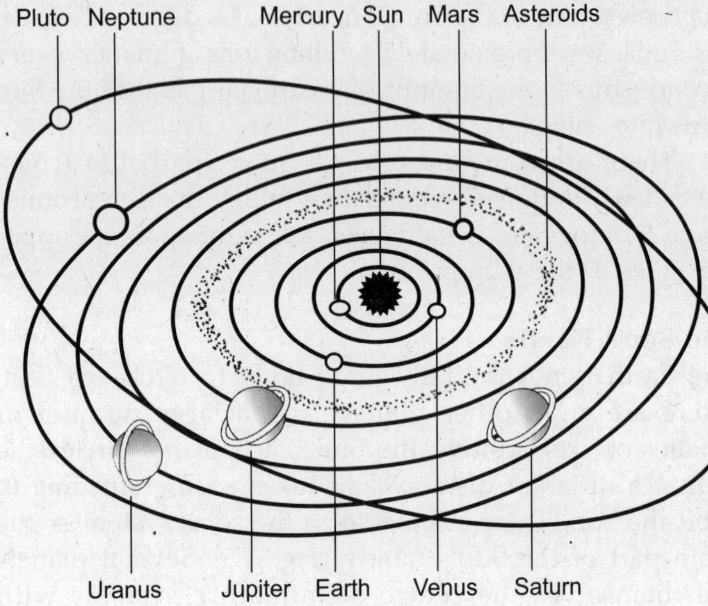

Fig. 1.5: Solar system

Physical Characteristics of the Sun

Diameter	1,390,000 km (863,706 miles)
Mean Distance from Earth	149,597,890 km (9.30 x107 miles) or 1 astronomical unit
Age of Sun	4.5–4.7 x 109 years
Volume (Earth=1)	1.412 x 1033
Mass	1.99 x 1030 kg (7.02 x 1031 ounces) or 332,830 (Earth = 1)
Density	1.41 gm/cm3
Surface gravity	28 (Earth = 1)
Length of day or Differential rotation	25.38 Earth days. At the equator the surface rotates once every 25.38 Earth days; near the poles it's as much as 36 Earth days.
Temperature	5,503.85° C (9,935.93° F)

Inside the Sun

Beneath the photosphere, or the surface of the Sun, are the violently churning convection zone, the radiative zone and the core, where the Sun's energy is produced (Fig. 1.6). This energy flows from the core to the photosphere and then out into the space as radiant heat and light.

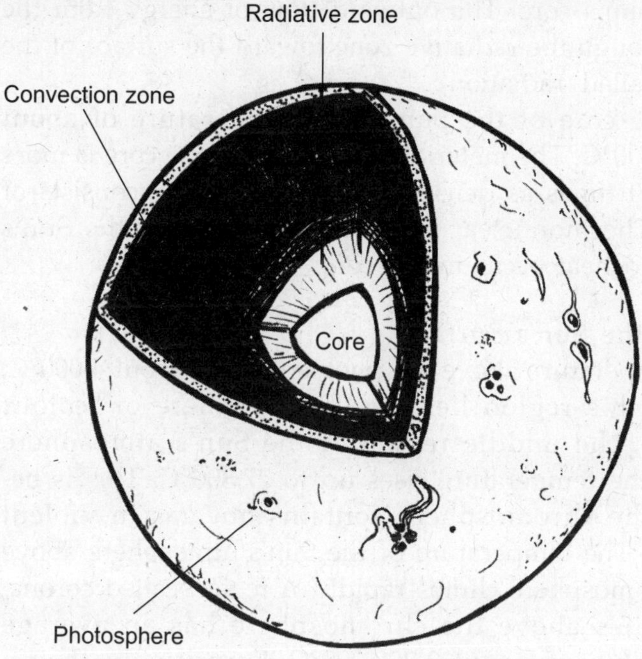

Fig. 1.6: Inside the Sun

The convection zone has a temperature of about 1,100,000° C and the gases here are about a tenth as dense as water. Due to the opaque nature of the gases, the energy from the Sun's core cannot travel through the convection zone by radiation. Therefore, the energy causes the gases to undergo violent churning motions or turbulence called 'convection', carrying most of the Sun's energy to the surface.

The radiative zone extends below the convection zone. The temperature here is much higher than the convection zone going up to 2,500,000°C and the gases are about as dense as water. The parts of the radiative zone that are nearer the Sun's surface are cooler than those that are closer to the Sun's core. The outward flow of energy from the core through the radiative zone toward the surface of the Sun is called 'radiation'.

The core of the Sun has a temperature of about 15,000,000°C. The material that makes up the core is more than 100 times as dense as water but it still consists of gases. Thermonuclear reactions that produce the Sun's light and heat occur in the core.

Above the Sun's Surface

The temperature above the photosphere is about 4,000°C. Above this region lies the chromosphere or 'colour sphere', the middle region of the Sun's atmosphere where the temperature rises up to 27,800°C. This is because the chromosphere contains hot gas in violent motion. The temperature of the Sun's atmosphere above the chromosphere climbs rapidly. A region called corona, which lies above the chromosphere has an average temperature of about 2,200,000°C. Surprisingly, the atoms of the corona are so far apart that the gases within the corona have little heat, so much so, that if an astronaut landed in the corona, his spacesuit would need heating. The corona has no well-defined boundary and its gases (solar wind) constantly expand away from the Sun. The temperature drops slowly from the corona outward into space.

This phenomenon is puzzling. Theoretically, heat flows from hot areas to cooler areas, and yet the photosphere is cooler than the outer regions of the Sun's atmosphere.

The Sun's Surface

The innermost layer of the Sun's atmosphere, which is about 550 km thick, is at a temperature of about 5,500°C. It is from one-millionth to 1 ten-millionth as dense as water. The photosphere contains many small patches of gas called 'granules'. These may have been formed by the violent churning of the gases. A granule may last for five to ten minutes and then fade away giving rise to new ones. The dark spots within the photosphere are called Sunspots.

The photosphere emits Sun's energy in the form of heat and light. Light given off by the photosphere is made up of many colours. Various elements in the photosphere absorb some of the colours and prevent the Sun from giving off those colours.

Activities of the Sun

At times the Sun turns 'active'. Spectacular events take place on Sun's surface and when these turn violent in nature, solar storms occur. Generally these appear above Sunspots in the chromosphere and the corona. Solar storms look like bright bursts of light called 'flares', which release a lot of solar energy. There are other solar storms that occur as huge arches of gas called 'prominences'.

Why do these activities occur on the Sun? Sunspots, flares, prominences etc. occur because of changes in the patterns of magnetic fields on the Sun. The Sun has a magnetic field that somewhat resembles a bar magnet, especially near its poles. This magnetic field is made irregular due to the movement of gases whose atoms are ionised. Many atoms on the surface of the Sun are ionised having lost electrons forming plasma. Particles trapped in a magnetic field usually follow the magnetic lines of force but the motion of large quantities of plasma tends to change the direction of these lines. It is this change in the pattern that gives rise to the Sun's activities.

The activities of the Sun have caused dramatic events that have gone into historical records. In 1940s, during the World War II, the British radar system suddenly went berserk inducing panic amongst the British soldiers who thought that the Germans were about to launch an air-raid. In the 1970s, Skylab, an American space station re-entered the Earth's atmosphere almost two years earlier than its scheduled time.

The 1980s saw many mishaps due to the 'active' Sun. A powerful explosion of a gas pipeline destroyed the Trans Siberian Railroad in the year 1989 causing the burning of two passenger trains, killing about 500 people. On 13 March 1989, Quebec, a province in Canada went into sudden darkness for almost about nine hours due to power failure. Around the same time, in British Columbia and Ontario, needles on recording devices at radio telescope facilities, went haywire.

In the late 1990s seven Iridium communication satellites and several other communication satellites failed.

Sunspots

Sunspots are the cooler areas of the Sun. These appear as irregular dark patches on the Sun's surface. Chinese astronomers spotted sunspots 2,000 years ago. In 1612, the Italian astronomer, Galileo traced the Sunspots through a telescope (Fig. 1.7). The spots glided across the Sun's face near the equator. Astronomers Thomas Harriot, Johannes Fabricius and Christopher Scheiner observed Sunspots through a telescope. Historical records show that Thomas Harriot was the first to observe Sunspots through his telescope in the year 1610. Johannes Fabricius then

Fig. 1.7: Galileo Galilei

went on to publish a book on Sunspots, *"De Maculis in Sole Observatis"* in 1611.

At times, a strong loop of magnetic lines of force extends through the Sun's surface lowering the temperature of the gas, which does not shine so brightly and appears as a Sunspot. A typical Sunspot may have a diameter of about 32,000 kilometres. Most Sunspots have two sections, the inner one called *umbra* with a diameter of 13,000 kilometres—as much as the size of Earth and the outer one called *penumbra,* having a diameter of 19,000 kilometres. The penumbra is hotter and brighter than the umbra. However, some small Sunspots have no penumbra.

The number of visible Sunspots varies from 0 to about 100. On an average it takes about 11 years for the number to increase from a minimum to a maximum and back to another minimum. This is called the Sunspot cycle. The Sun actually takes two Sunspot cycles or about 22 years to go through a complete set of magnetic changes. During the maximum phase of a Sunspot cycle, all types of solar activity become very intense. It was Samuel Heinrich Schwabe who in 1843, discovered this cyclical increase and decrease in the number of Sunspots with time. At present, we are in the late 23rd cycle of the Sunspot cycle, the solar maximum of which occurred in the year 2000. The cycle will last until the year 2006.

According to scientists, the Sunspot activity on the Sun is responsible for the condition in the Pacific Ocean called *El Nino. El Nino* is a substantial increase in the water temperature that results in an increase in the number of storms that occur. It also affects other areas of weather such as the amount of rainfall or snowfall that occurs. The weather condition called *La Nina* has also been connected to Sunspot activity.

Flares

The magnetic lines of force get jumbled at times when the Sunspot group exists for a long time. The magnetic energy then gets stored in the corona and the energy gets released in a discharge called a flare. The energy is released in the form of light, heat, fast-moving atomic nuclei and electrons called solar cosmic rays. During a coronal mass ejection or a solar flare, millions of tons of material is thrown out into space with a speed as high as 1,000 kilometres per second. Some of this material reaches Earth within 2 to 3 days and these induce severe magnetic storms impacting satellites and disrupting communication.

Large flares disrupt radio communications due to production of excess cosmic rays. These endanger astronauts in space, where the Earth's magnetic field is not present to protect them from such large amount of radiation. Eruption of solar flare may even be accompanied by a display of auroras, the dramatic and beautiful display of colours in the sky.

Prominences

Gases in prominences have a higher density and they radiate light more efficiently than do gases in chromosphere and the corona. Prominences are of two kinds—quiescent and active. An active prominence changes rapidly during a period of several hours erupting and flinging gases rapidly into space, contrary to the quiescent type, which changes little in appearance during its two to three month existence.

Solar Radiation

Besides emitting visible light and heat, the Sun gives off many kinds of radiation such as radio waves, ultraviolet rays and X-rays. Strong bursts of radio waves occur during violent solar activity originating in the Sun's atmosphere above the Sunspot groups, chiefly when flares occur. Ultraviolet rays are the invisible rays comprising of waves of light shorter than the waves of violet light on the visible spectrum. These are the rays that cause sunburns, and excess exposure may even cause skin cancer. During the periods of violent activity, the Sun gives off more ultraviolet rays and X-rays than during calm periods. X-rays are capable of destroying the tissues of living creatures but the Earth's atmosphere shields us from most of the solar radiation.

Solar Wind

The solar wind is the flow of gases coming from the regions of hot corona that has low temperature and density, called coronal holes. The solar wind confines the Earth's magnetic field into a specific volume of space called the magnetosphere. Sometimes flares inject high-speed particles into the solar wind. The solar wind presses harder on the Earth's magnetosphere causing magnetic storms on Earth. These magnetic storms interfere with radio communications and they can make the compass needles swing wildly. The solar wind has large effects on the tails of comets and even has measurable effects on the trajectories of spacecraft.

Recent data from the spacecraft Ulysses show that during the minimum of the solar cycle the solar wind emanating from the polar-regions flows at nearly double the rate, 750 kilometres per second, than it does at lower latitudes. The composition of the solar wind also appears to differ in the polar-regions. During the solar maximum, however, the solar wind moves at an intermediate speed. The recently launched Wind, ACE, and Solar and Heliospheric Observatory (SOHO) spacecraft will further study solar wind from the dynamically stable vantage point directly between the Earth and the Sun about 1.6 million kilometres from Earth.

The Sun's Rotation

The Sun has a very strong magnetic field that is thought to be responsible for flares and Sunspots on its surface. The magnetic activity of the Sun is driven by differences of rotation rate in the Sun. At the equator, the period is about 25 days while at the poles the rotation period is around 31 days. The amount of solar activity varies through a 22-year cycle. The time when the solar activity is at a maximum, it is called a "solar maximum".

Sun Study

The SOHO project is being carried out by the European Space Agency (ESA) and the US National Aeronautics and Space Administration (NASA) as a cooperative effort between the two agencies in the framework of the Solar Terrestrial Science Program (STSP) comprising SOHO and CLUSTER, and the International Solar-Terrestrial Physics Program (ISTP), with Geotail (ISAS-Japan), Wind, and Polar. SOHO was launched on 12 December 1995. The Observatory is primarily launched to study the internal structure of the Sun and its atmosphere and the solar wind.

The SOHO spacecraft was built in Europe by an industry team from Matra-Marconi Space (MMS) Company. European and American scientists provided instruments. There were nine European and three American Principal Investigators. NASA was responsible for the launch and is now responsible for mission operations. Large radio

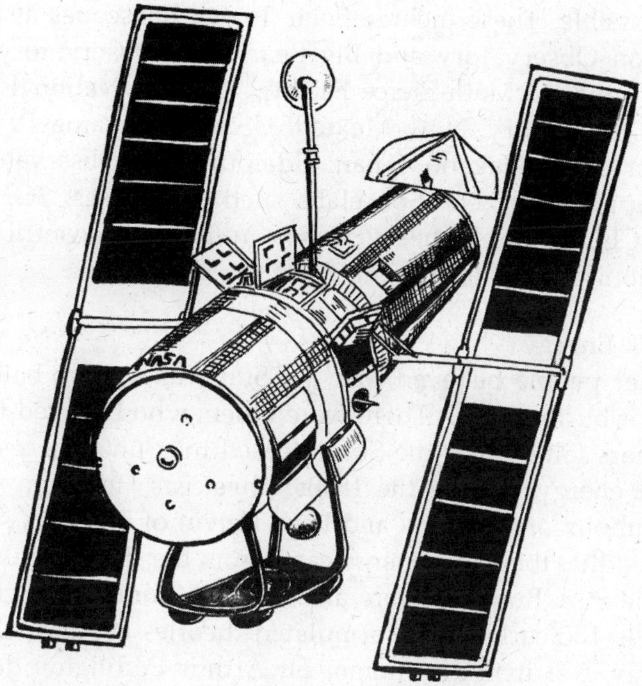

Fig. 1.8: Hubble Space Telescope

dishes around the world that form NASA's Deep Space Network are used to track the spacecraft beyond the Earth's orbit. Mission control is based at Goddard Space Flight Centre in Maryland (USA). Incidentally, Hubble Space Telescope (Fig. 1.8) project also originated from the Goddard Space Flight Centre. Hubble orbits 600 km above Earth, working round the clock to unlock the secrets of the Universe.

SOHO uses 12 different types of imaging instruments. It provides a view of the Sun from space. Unlike the other satellites, SOHO is designed in such a way that one can have a continuous viewing of the Sun.

Several ground-based facilities help in observing the Sun and collecting as much information about the Sun as possible. These include Solar Tower Telescopes at Mt. Wilson Observatory and Big Bear Solar Observatory at California; McMath-Pierce Facility, Arizona; National Solar Observatory, New Mexico, USA; Nobeyama Radio Observatory, Nagano, Japan; Udaipur Solar Observatory, Udaipur, India; etc. Spacelabs such as Ulysses, SOHO, TRACE, Skylab, Pioneers, Helios, Yohkoh—all contribute to monitoring information about the Sun.

Sun's Energy

Earlier people believed that the Sun was a giant ball of fire or burning coal. There were others who believed that meteors falling into the Sun released the Sun's energy or solar energy. During the 1800s, physicists Hermann von Helmholtz of Germany and Lord Kelvin of England, suggested that the Sun's energy came from the slow shrinking of the Sun. But they were all wrong. A number of scientists in the early 1900s formulated theories about nuclear energy. A British astronomer Sir Arthur Eddington demonstrated that the centre of the Sun has a temperature of many millions of degrees due to which the nuclei of atoms unite in the process of thermonuclear fission. Physicists Hans Bethe of the USA and Carl F von Weizsacker of Germany described this process in the 1930s.

Theories Explaining Sun's Energy

The Sun rotates west to east on its axis every 25 to 34 days, and it revolves around the centre of our galaxy in approximately 230 million years. Since the Sun is not a solid body, all of its parts do not rotate at the same rate. It, however, is rushing through space with tremendous speed. The planets revolve around the Sun, held in place by gravity. Planets receive light, heat, and all the other electromagnetic radiations from the Sun, which is considered the main source of energy in the Solar System.

The Sun radiates energy into the space in the form of light and heat. And every second about 3.6 million metric tons of the Sun's mass changes into energy. But the Earth gets only about one two-billionth of the total energy radiated by Sun every second. And yet just this little of the Sun's energy is sufficient to make life possible on the Earth.

How is this tremendous amount of energy released from the core? The source of the Sun's fuel is hydrogen and helium gases. Hydrogen undergoes nuclear fusion releasing an enormous amount of energy in the form of light and heat. Two isotopes of hydrogen (tritium and deuterium) collide with each other under extreme heat in the interior of the Sun. The two atoms smash into each other at high speed—so hard indeed that several things happen.

Protons or two hydrogen nuclei smash into each other to fuse forming a deuterium nucleus with one proton and one neutron. As the protons fuse, they release energy partly in the form of neutrinos. The deuterium nucleus then rams another proton, fuses with it giving off energy in the form of gamma rays. Finally, the clumps of three particles smash into each other forming a helium nucleus and during this collision, two protons are liberated; the process begins again. Each of these processes gives off energy in the form of protons. In just one second, the Sun emits more energy than all humans have ever produced during their entire stay on Earth. Yet our planet receives only about two-billionth of the total energy output of the Sun; the rest spreads out in all directions of the space.

The Sun Will Not Shine Forever

Like us, the Sun was born and shall one day die. About five billion years from now, scientists believe that the centre of the Sun will shrink and become hotter. The surface temperature will fall slightly while the higher temperature at the centre will increase at the rate at which hydrogen changes into helium. The amount of energy given off by the Sun will also increase. The Sun will then start to expand, swell to become a red giant star, 166 times its present size, and an inflatable hot air atomic oven, which will radiate hugely increased heat. The ice caps on Earth will melt, the oceans will evaporate, and the deserts will prevail. Due to extreme heat, life will cease to prevail on Earth. Venus will sizzle and Mercury will be cooked to a crisp.

Once the Sun has used up all its thermonuclear energy, astronomers believe that it will begin to shrink to about the size of the Earth and will turn into a 'white dwarf'. Ultimately, that solar forge, source of all energy, will slip into gradual decline and shrink down to what is called a cinder of its former fiery self. It will be a compressed black coal at its burnout 50 billion years from now, frizzled in its own furnace. The white dwarf would have used up all its energy and lost all its heat turning into a cold, dark globe. Stone cold dead! Such stars are called 'black dwarfs'. The planets will then become cold and dark. Earth would be vastly changed and devoid of life as we know it. The gases of the atmosphere would freeze onto the Earth's surface.

Travel to the Sun

Our Sun is a one million kilometre-wide ball of burning gas. So be prepared for the best fireworks you have ever seen! A journey to the Sun itself would be impossible even for the best protected of ships due to the scorching temperatures. And yet in case you plan a trip to the Sun, before you leave make sure you obtain a timetable of Sunspot activity because these are the cool spots. Anyone attempting to fly close to the Sun will have to be aware of the solar wind. It will have a sig-nificant effect on the path of any craft approaching the Sun.

Missions to the Sun

Ulysses: This is a joint project between NASA and the European Space Agency (ESA). The *Ulysses* spacecraft is orbiting the Sun to study its north and south poles.

Launch Date: 6 October 1990

First South Pole Pass: 1994

First North Pole Pass: 1995

The solar system's deflector shield is a giant magnetic bubble called the 'heliosphere', which is a part of the Sun's magnetic field. A stream of neutral helium atoms is flowing into the solar system right now. *Ulysses* has been sampling the stream itself, snatching atoms directly from the flow.

The stream is also actively monitored by NASA and ESA spacecraft SOHO, EUVE and ACE.

Genesis: The *Genesis* spacecraft was placed into orbit around L1, a point between Earth and the Sun where the gravity of both bodies is balanced, to collect particles of the solar wind.

Launch Date: 8 August 2001

Genesis returned to Earth on 8 September 2004. But its parachute system failed to deploy. Scientists at NASA's Johnson Space Centre (JSC), Houston, have been unpacking samples of atoms and ions from the solar wind collected during the nearly 3-year mission of *Genesis* in deep space. Planning is underway for preliminary examination of the samples.

2

NATURAL EVENTS

If the Sun simply stopped emitting its energy into space, the Earth and all planets would be plunged into eternal darkness. The only light we would be able to see would be the faint light from the distant stars as all the planets; their satellites, asteroids, and comets shine by reflecting Sunlight. We would all be frozen solid without the Sun. We are therefore totally dependent upon the Sun for our survival. Without the Sun plants cannot grow. Without plants there would not be food for animals. Soon no life would be left.

A casual observer may have little interest in the Sun and, in general, he or she has a tendency to take the Sun for granted. The Sun appears to be a ball of bright light, much too bright to be looked at directly. During the summer months, we notice that the Sun is high in the sky at noontime. In the winter, the Sun is low in the sky. Few people realise that the Sun always travels across the sky from east to west, or that for all latitudes north of the Tropic of Cancer the Sun never appears to be overhead at noon. The Sun can never be seen overhead at noon in the Southern Hemisphere for all latitudes greater than the Tropic of Capricorn.

There has always been a desire to understand the reasons for the various phenomena of nature, which are observed in the sky. Early in mankind's existence on Earth, it was recognised that only one visible object—the Sun—controlled life.

Who does not love to admire a sunrise, a sunset or the appearance of a rainbow in the sky? Be it the rising Sun or the setting Sun or a rainbow—all involve a spectacular display of colours in the sky. These visuals in the sky are called 'atmospheric optics'. As sunlight enters the atmosphere, it is absorbed, reflected, scattered, refracted or diffracted by atmospheric particles or air molecules. These processes, individually or in combination, are responsible for producing most optical effects. The type of optical effect that results greatly depends upon the type of particles the light encounters and on the wavelength of the light.

Why Does the Sun Shine?

The nuclear reactions between hydrogen atoms within the Sun's core produce a tremendous amount of energy making the Sun shine. Photons and other kinds of particles carry the energy to the Sun's surface. At the surface, these particles excite atoms and then de-excite by emitting light. The Earth receives a part of that light. Not only the Sun, but all the stars also emit light by this process.

Sunlight

Light from the Sun is the natural source of light that heats the Earth and makes life on Earth possible. Plants need sunlight to grow. Animals eat plants or other animals that have been nourished by plants. Directly or indirectly, all life depends on the Sun. Without the Sun's light, the Earth would turn so cold and frozen that nothing could live on it. Even if we burned all our fuels, we may not be able to keep the Earth warm enough for life to exist.

Sunlight is made of tiny units of radiant energy called photons. The photons are born in the inferno of the Sun's core. Depending on the amount of energy a photon has, it may be absorbed by the Earth's atmosphere. Although each photon carries only a speck of energy, trillions of photons hit each square meter of Earth every second. Together, they form sunlight. Clouds, water and ground, all reflect sunlight in the space, which allows us to see Earth from a spaceship. The amount of energy that is carried by the wave of light or a photon largely determines the colour of the light.

Light is an electromagnetic wave as it consists of both electric as well as magnetic fields. For light to be visible, it must have a wavelength within a certain narrow range of values called the visible spectrum. For instance, violet light has the shortest wavelength and hence it is visible; while red light has the longest wavelength. Between these lie all the other colours each with differing wavelengths. When seen together at the same time, the colours appear as white. Sunlight is white because it has all the colours. Only when it passes through a prism, do the colours separate out.

Sunlight spread into its different colours by a prism, creates a continuous spectrum. From violet to red it blends smoothly from one colour to the next. When we look at the Sun through a spectroscope, it is like looking at light from the bulb shining through a gas. When fitted to a telescope, a spectroscope acts like a prism. It spreads the light out in the rainbow band called a spectrum. The dark-line spectrum shows the fingerprint patterns of hydrogen, helium, iron, magnesium and about 70 other elements of the chromosphere and corona. Each type of atom in the Sun's atmosphere absorbs certain colours. When we see the Sun's light through a spectrometer, we can see that the sunlight has thousands of gaps. This is because the light source has travelled through a gas that absorbed certain colours.

The visible spectrum is only a part of the electromagnetic waves. Those waves with slightly short wavelengths are called the ultra-violet rays. Waves, which are still shorter than the latter are the X-rays while those that are yet shorter, are the gamma rays, which result from the nuclear reactions of the Sun. Those waves that are slightly longer than those of the red light are called infra-red rays. When you stand in bright sunlight or in front of a fire, you feel warm because of the infra-red rays shining on you. Radio waves are those that have still longer wavelengths than infra-red rays.

Why is the Sky Blue? Blame it on the Sun's light!

On a sunny day, the sky looks blue. Sometimes, early in the morning and just before sunset, it looks greenish-yellow or even orange-red. The sky actually has no colour at all. It is a great ocean of air called the atmosphere made up of colourless gases. The sky looks coloured to us because of what happens to the sunlight, which passes through it.

Particles such as dust, smoke, ash etc. float in the air. Burning fuels, strong winds that whip up particles of sand, cloud and smoke from volcanic eruptions, all these give out particles that float in the air. When the Sun is high in the sky, its light is scattered by the particles in the atmosphere. The light photons with the highest energy—violet and blue are scattered more readily than those with a lower energy— red and orange. The Earth's atmosphere filters out all the other colours of the light spectrum in the Sunlight. So, the only colour left is blue, which is what's scattered throughout our atmosphere and reflected in the sky.

Sunrise and sunset are the most beautiful sights that even artists and painters clamour for. Each one of us may have craved to see a sunset or a sunrise at some time or other. For this magnificent miracle of nature people climb hills and mountains or go on a stroll on beaches. When objects such as mountain peaks or clouds partially shadow the Sun's rays, crepuscular rays occur. The name crepuscular means, "relating to twilight" and these rays are observed at sunrise and sunset. Crepuscular rays appear to diverge outward from the setting Sun, and are visible only when the atmosphere contains enough haze or dust particles so that Sunlight in unshadowed areas can be scattered toward the observer. The light rays are actually parallel, but appear to converge to the Sun due to 'perspective', the same visual effect that makes parallel railroad tracks appear to converge in the distance. Crepuscular rays are often red or yellow in appearance because blue light from the Sun is selectively scattered out of the beam by air molecules.

Light rays scattered by dust and haze occasionally appear to converge toward the "antisolar" point, (the location on the horizon opposite the point where the Sun is setting). These rays, called anti-crepuscular rays, originate at the Sun, cross over the sky to the opposite horizon, and appear to converge toward the antisolar point because of perspective.

Twilight

At sunrise and sunset, we see the Sun through a greater thickness of air because it is low in the sky. Therefore, we are looking at it across the Earth's surface rather than directly up into the sky. As the Sun is setting, its light travels to us across the Earth's surface. The scattering of Sunlight by the upper layers of the Earth's atmosphere causes the great yellow and orange hues that we see in the evening sky. This is because the photons of red and orange light are scattered through the atmosphere around us.

There are different definitions of twilight. Astronomical twilight begins at sunset (ends at sunrise) and ends (begins) when the Sun's centre is 18 degrees below the horizon. While our twilight begins (ends) when the Sun's centre is 6 degrees below the horizon, nautical twilight begins (ends) when the Sun's centre is 12 degrees below the horizon.

Aurora

If you live in the extreme northern or southern regions of the world, you might see an aurora—a dazzling display of coloured lights—green, blue or red that flicker in the sky at night. When large eruptions take place on the Sun, high-speed particles arrive at the Earth and go crashing into the air molecules. These collisions excite the molecules releasing extra energy in the form of light. When the collisions occur at night, the light emitted may be bright enough to be seen. An aurora, such as the northern lights, is an emission of light by molecules of air.

The oxygen in the Earth's higher atmosphere creates red auroras, while in the lower atmosphere it creates green auroras. Atmospheric nitrogen gives rise to the blue and red auroras. These may change in less than a minute. The probability of the display of an aurora increases with an 'active' Sun.

Day and Night

The planet Earth travels through space around the Sun and it also spins on its own axis. A solar day is the length of time that it takes the Earth to turn around once with respect to the Sun. Babylonians began their day at sunrise while the ancient Jews began their day at sunset. The Egyptians and Romans were the first to begin their day at midnight. For us, a day begins when we start for work and ends when we get back home and get a good night's sleep. According to us, when the Sun is shining, it is day, and the night falls when there is no Sun. Each day for us begins at midnight, when the calendar day changes and is divided into two parts of 12 hours each. The hours from midnight to noon are the a.m. (ante meridian) or before noon hours. Those from noon to midnight are the p.m. (post meridian) or after noon hours. And this is depicted very clearly on our time-keeping devices according to which we work.

The length of the daylight hours changes during the year in all parts of the world. This is because the tilt of the Earth's axis causes the North Pole to face the Sun and then the South Pole as the planet Earth orbits the Sun. The longest day in the Northern hemisphere usually is June 21 and the shortest day is December 21, while the longest day in the Southern hemisphere is December 21 and the shortest day is June 21. Each of these long days has 13 hours and 13 minutes of day at 20-degree latitude and this changes as the degrees in latitude increases. The shortest days consist of 10 hours and 47 minutes of day-light at 20-degree latitude, which becomes still shorter as the degrees in latitude increase (Fig. 2.1).

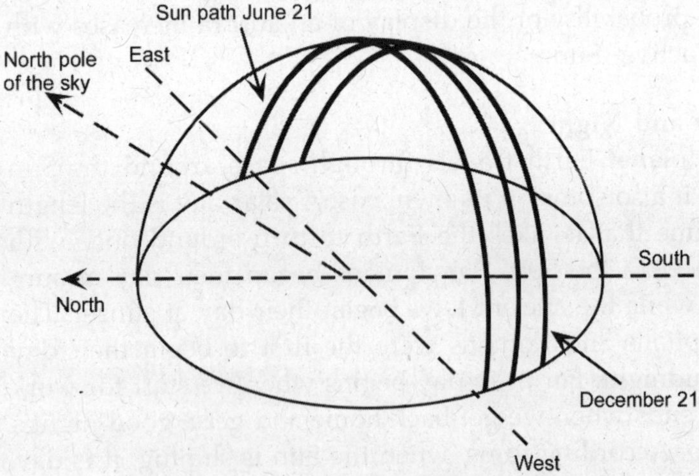

Fig. 2.1: The longest and the shortest day

Equinox

For two days of the year, the Sun is directly above the Earth's equator. At these times the days and nights are of nearly equal length everywhere on the Earth. These are called equinoxes, which occur on March 20 or 21 and on September 22 or 23. The time difference between the two equinoxes is not the same because of the Earth's elliptical orbit.

Temperature on the Earth

When we think of temperature we think that either it is too hot or too cold. The relationship between the Sun and Earth's temperatures is very close as the activity on the Sun plays a big role in what temperatures will be on Earth. Why is it hotter at the equator than in the Arctic? Or why is the Arctic 'warmer' than Antarctica? It is all directly related to their relative location to the Sun.

The Earth, like the Sun gives off energy in the form of light (radiation) mostly at invisible infra-red wavelengths. The amount of Sunlight that the Earth absorbs is an important factor in determining its temperature. The hotter the Earth gets, the more energy it radiates. If the Earth were suddenly to absorb less Sunlight, it would cool until the energy arriving and leaving the planet were in balance again.

The Earth is in such an exact position that the average temperature is balanced between the freezing point of water (0°C) and the normal body temperature of humans (37.8°C). If the average temperature ever varied very far above or below these two extremes for very long, we would not survive. The rotation of the Earth on its axis causes us to have alternate periods of day and night. During the day, the side toward the Sun warms up while the other side cools down. This has a big effect on the Earth's average temperature.

If our Earth was even just 5 per cent closer to the Sun, the increase in the temperature would be enough to melt the ice caps at the North and South poles. The water trapped in these caps of ice would then raise the oceans level by as much as 9,000 centimetres. That would be enough to submerge every coastal city in the world.

Temperature Is On the Rise

Over the past few years, human activity has caused the Earth's surface temperature to rise. Sunspots on the Sun results in radiation activity in the Sun and this in turn, raises the temperature on the Earth. Scientists suggest that since the 18th century, the number of Sunspots on the Sun have increased while the Earth is cooling down very slowly. Presence of pollutants such as carbon dioxide, methane, sulphur dioxide and other gases in the atmosphere also contribute to the increase in the greenhouse effect.

In the last century, the temperature increased by 0.5 to 1 degree, while predictions of climate change in this century due to human activity range anywhere from 2 degrees to 6 degrees. This will result in a change in the global climate, i.e., there will be a change in the long-term weather patterns in different regions of the world. (The term 'weather' refers to the daily changes in temperature, wind, and/or precipitation of a region.) The Sun heats the Earth's atmosphere and causes air and water to move around the planet. The result can be a slight breeze or a tornado.

Global Warming

The temperature on Earth rises due to the carbon dioxide (CO_2) being trapped in the atmosphere. Scientists suggest that by burning excessive amounts of fossil fuels, we have released extra CO_2 molecules and increased its percentage in the composition of the atmosphere. The CO_2 concentration has increased from 280 ppm to 360 ppm since the Industrial Age (Fig. 2.2).

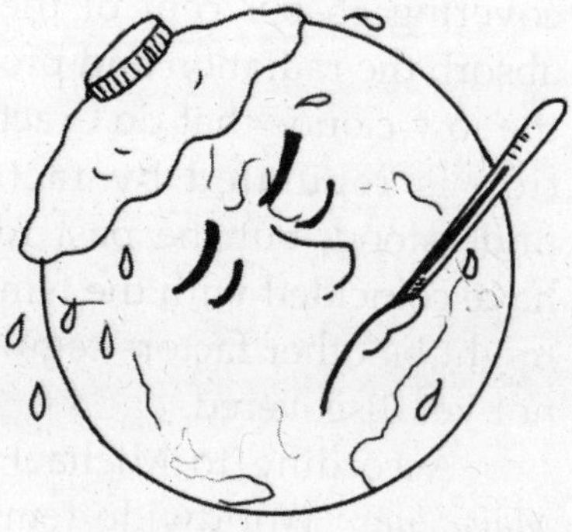

Fig. 2.2: Global warming

Increasing concentrations of water vapour (H_2O), gases such as carbon dioxide (CO_2), methane (CH_4), Chloro Fluoro Carbons (CFCs), and ozone also increases the greenhouse effect, trapping additional heat. This leads to global warming. Some of the molecules absorb more heat than others. For example, a CFC molecule can retain heat up to 1500 times more than a carbon dioxide molecule. The USA accounts for well over 25 per cent of all the CO_2 emitted worldwide, and that percentage is steadily growing. To make matters worse, many other populous developing nations such as China and India are also rapidly expanding their emissions.

More CO_2—More Heat

A few skeptics believe that global warming is not caused by human activities. Instead, they suggest that most of the warming is caused by other factors. For instance, the Sun undergoes periods of cycles every 22 years, during which time, its cosmic ray activity varies from high to low. When the solar activity is plotted on the graph with the Earth's temperature, there is strong direct correlation. As the solar activity increases, the temperature of the Earth also increases, and *vice versa*.

Another reason for global warming is speculated to be the cloud formation. There are two types of clouds covering 65 per cent of the Earth—the high clouds that absorb the radiation and provide additional warming, and the low clouds that do exactly the opposite. Cloud formation is regulated by factors that are still not quite understood, but the past records of the cloud formation have coincided with the temperature variation. Also, there might be other factors regulating the temperature that are not yet discovered.

According to Michael Lemonich (April, 2001, *Time Magazine*), 'Worldwide temperatures have climbed more than one degree Fahrenheit over the past century, and the 1990s were the hottest decade on record! Mt Kilimanjaro has lost 75 per cent of its ice cap since 1912. The ice on Africa's tallest peak could vanish entirely within 15 years. In Eastern Siberia Lake Baikal freezes 11 days later than it did a century ago. The Arctic permafrost is starting to melt! Mountains will lose all their glaciers in Glacier National Park by 2070 if their retreat continues at the current rate! Venezuelan mountain tops had six glaciers in 1972. Today, only two remain. Coral reefs are dying off, as seas get too warm for comfort. Plants and animals are shifting their ranges pole ward and to higher altitudes. Migration patterns for animals as diverse as polar bears, butterflies, and beluga whales are being disrupted'.

Recent predictions indicate that the average world temperature will increase by 2.5 to 10.5 degrees Fahrenheit by 2100, and create a much worse situation than previously estimated.

Climate and Sun—Hand in Hand

Because of the spherical shape of the Earth, the energy of the Sun does not reach all the parts of the Earth to the same extent. The equator receives more direct heat and light because the Sun is directly overhead. This makes the climate hotter near the equator.

At the poles, the Sun is not directly overhead; it is always near the horizon. As a result, the polar regions receive less heat, so the climate there is cold. The white ice of the poles also tends to reflect the heat energy, which makes it even colder. On the other hand, the green plants of the tropics tend to absorb more heat, which makes it a warmer region.

The Four Seasons

We know that the Earth spins around on its axis once a day. The Earth also travels in an elliptical orbit around the Sun. It takes just over 365 days to make that journey. This is the period we call a year. In a year, we have different seasons. The seasons on Earth are a visible reminder of the way that the Sun affects our lives. Generally, seasons include periods of the year consisting of spring, summer, autumn and winter, lasting for about three months each. These bring in changes in temperature, weather and length of daylight.

Some places on our planet think in terms of two seasons only—rainy and dry. At the polar regions the seasons are different yet again, due to their being located near the poles. Regardless of where you live and how you think about the seasons, the Sun is the reason why we have these variations (Fig. 2.3).

The seasons change because places on Earth receive different amounts of sunlight during the year. The tilt of the Earth's axis at an angle of about 23.5 degrees affects the amount of sunlight received by the different regions. When the North Pole is tilted toward the Sun, the Northern Hemisphere enjoys summer and six months later, when the axis is pointing away from the Sun, the Southern Hemisphere enjoys summer. The Sun's rays strike the Earth from a high angle and each place receives maximum Sunlight. When the North Pole is tilted away from

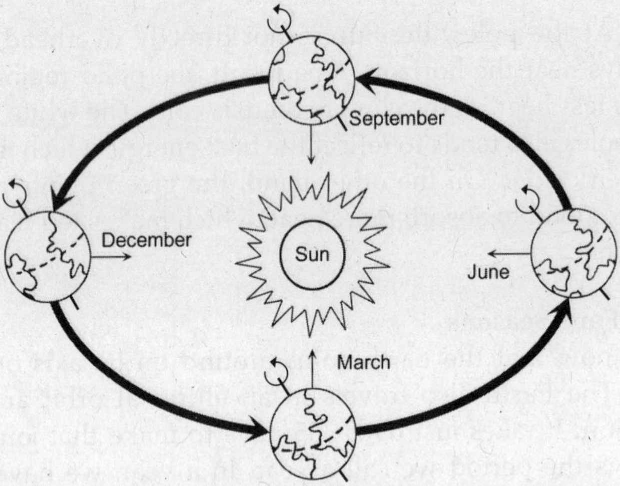

Fig. 2.3: Seasons

the Sun, the Northern hemisphere has winter. The Sun's rays come from a lower angle and each place receives minimum Sunlight. As the Earth moves between these positions, autumn and spring occur.

The Sun traces out an arc as it travels. In the Northern hemisphere as winter gets gradually nearer, the Sun's path shifts southward to a low arc, which makes the days shorter and the nights longer. As summer approaches the Sun's path grows higher causing longer days and shorter nights. The results of this pattern are reversed for the Southern hemisphere.

In many parts of Asia, say in India, life revolves around the monsoon or the winds that bring rain. This marks the rainy season. Winds arise from low pressure areas created by the Sun's heat in tropical regions. The surrounding air moves into low pressure areas, and this movement produces winds. Saturated with moisture, the air rises and flows toward the polar region. The moisture condenses in the cooler upper air and is released as rain.

Sun and Air

We breathe in air. Air moves the clouds. Air helps create waves. Wind pollinates some plants. Without air, life on Earth would not exist. And without Sun the air would not have moved. Without the Sun we would not get oxygen because it is the Sun which initiates photosynthesis, which provides oxygen for us to breathe.

The Sun warms up the oceans and the air around them, thus creating wind. When the ocean currents and wind stir the ocean, it moves the ocean water. This process keeps the water and air on the earth fresh, and prevents it from stagnating.

The Sun and the moon also cause air tides. Just as the moon and the Sun affect the water of the oceans, they also affect the air of the Earth's atmosphere. Gravitational attraction causes air to form large bulges (air tides) just like water tides. These are also called lunar winds. Even though they are too slight to be felt, lunar winds cause dramatic variations in weather conditions.

What Causes Rain?

The heat of the Sun causes evaporation from the water bodies forming water vapour. The warm and moist air cools as it rises and the amount of water vapour it can hold, decreases. The temperature at which air holds as much vapour as it can is called 'dew point'. When the temperature drops below the dew point, some of the water vapour condenses into water droplets forming clouds.

These water droplets form on tiny particles of matter consisting of dust, salt from ocean sprays or chemicals from industries or vehicles, known as the 'condensation nuclei'. As the water droplets form, heat is released, making the clouds warmer. The warmth helps the clouds rise, and then they become cooler (Fig. 2.4).

Weather experts give two explanations for the

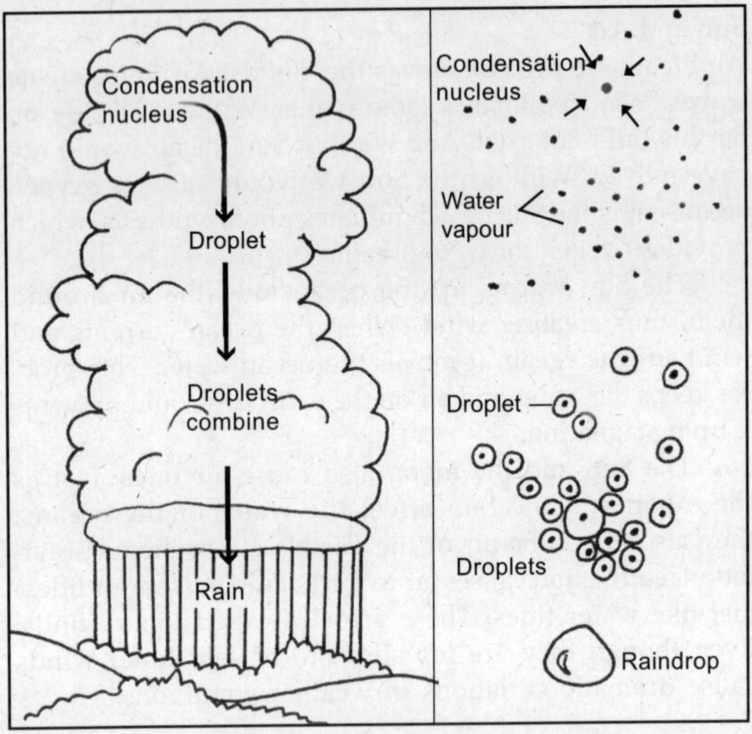

Fig. 2.4: How rain forms

formation of rain. One is the *coalescence theory*, in which the water droplets formed on the condensation nucleus fall through a cloud and combine with smaller ones. When these become too heavy for the air to support, they fall as rain-drops. The other explanation is the *ice-crystal theory* that involves the super-cooled water droplets. Such a droplet freezes on a particle called an ice nucleus. This forms an ice crystal, which combines with other crystals to form a snowflake. When the snowflake falls through air warmer than zero degrees, it turns into rain (Fig. 2.5).

Sometimes it rains even when it is sunny and there are no clouds in the sky. We call this a 'Sun-shower'. Strong winds bring the rain from clouds that are far away.

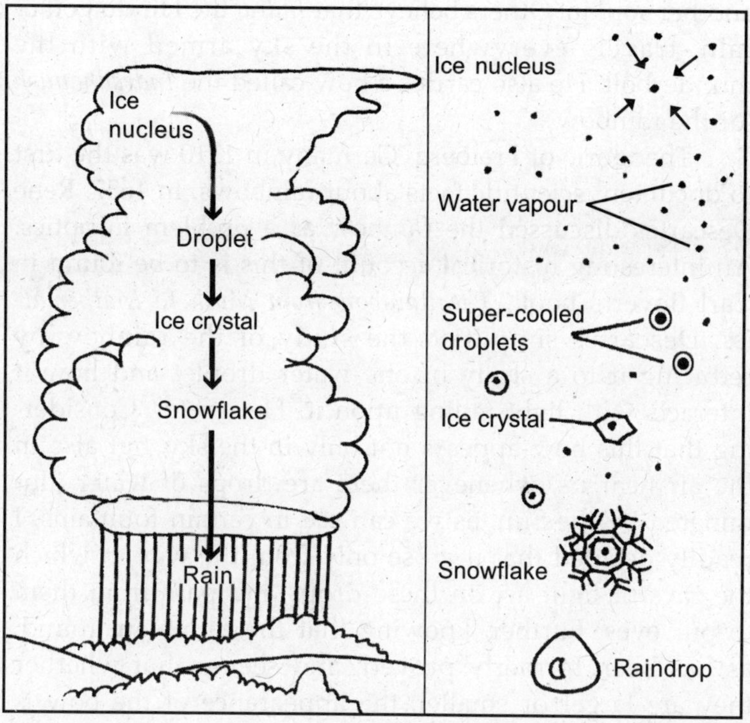

Fig. 2.5: The ice-crystal theory

Sometimes a Sun-shower is rain that falls from very high clouds. The clouds disappear before the rain can reach the ground.

The Magical Rainbow

There is no doubt that a sense of magic and beauty strikes one whenever a rainbow appears in the sky. Author Donald Ahrens in his text *Meteorology Today*, describes a rainbow as 'One of the most spectacular light shows observed on earth'. It looks like a giant celestial bridge and is called the 'Gateway to Heaven'. Many people believe that the rainbow is a ray of light falling on the Earth whenever Saint Peter opens the gates of heaven to let another soul in. Others believe that *Indra,* the Hindu god of rain, travels everywhere in the sky armed with the thunderbolt. He also carries a bow called the *Indradhanush* (or the rainbow).

Theodoric of Freiberg, Germany in 1310 was the first to document scientific facts about rainbows. In 1637, Rene Descartes discussed the rainbow as a problem in optics. An interesting historical account of this is to be found in Carl Boyer's book, *The Rainbow from Myth to Mathematics*. Descartes simplified the study of the rainbow by reducing it to a study of one water droplet and how it interacts with light falling upon it. He wrote: 'Considering that this bow appears not only in the sky, but also in the air near us, whenever there are drops of water illuminated by the Sun, as we can see in certain fountains, I readily decided that it arose only from the way in which the rays of light act on these drops and pass from them to our eyes. Further, knowing that the drops are round, as has been formerly proved, and seeing that whether they are larger or smaller, the appearance of the bow is not changed in any way, I had the idea of making a very large one, so that I could examine it better'.

A rainbow is Sunlight spread out into its spectrum of colours and diverted to the eye of the observer by water droplets (Fig. 2.6). The 'bow' part of the word describes the fact that the rainbow is a group of nearly circular arcs of colour all having a common centre. The colours of the rainbow arise from two basic facts—Sunlight is made up of the whole range of colours that the eye can detect. The range of Sunlight colours, when combined, looks white to the eye. Sir Isaac Newton first demonstrated this property of Sunlight in 1666. Lights of different colours are differentially refracted while passing from one medium (air, for example) into another (water or glass, for example).

Fig. 2.6: Rainbow

How Does a Rainbow Form in the Sky?

A rainbow is created by the Sun's interaction with water. It is formed when raindrops reflect Sunlight toward you, dispersing the light into colours. When the Sun shines during or shortly after a shower of rain, a rainbow forms in the part of the sky opposite the Sun. To see a rainbow, therefore, you must be standing with the Sun at your back and the rain must be falling in another part of the sky. The reflection, refraction and diffraction of the Sun's rays as they fall on the rain droplets cause this fascinating natural phenomenon.

It is interesting how a rainbow forms in the sky. As a ray of light passes through a drop of rain it is refracted (bent) and then diffracted (dispersed) into different colours. As the ray strikes the inner surface of the drop, it is reflected (turned back). And as it leaves the drop, it is further refracted and dispersed. Raindrops act as tiny prisms and mirrors to break up Sunlight into colours of the spectrum. Each of many raindrops sends out a coloured light at certain angles to form a rainbow. A rainbow shows seven colours—violet, indigo, blue, green, yellow, orange and red (VIBGYOR). The width and size of each coloured band varies depending upon the size of the raindrop. Larger drops cause narrow bands. Rays that reach the eye at a certain angle form each colour and the angle for a particular colour never changes.

Two Rainbows in the Sky

Sometimes we see two rainbows simultaneously present in the sky. This happens when we follow the path of a ray of sunlight as it enters and is reflected inside the raindrop. But not all of the energy of the ray escapes the raindrop after it is reflected once. A part of the ray is reflected again and travels along inside the drop to emerge from the drop. The rainbow we normally see is called the primary rainbow and is produced by one internal reflection; the secondary rainbow arises from two internal reflections and the rays exit the drop at an angle of 50 degrees rather than the 42 degrees for the red primary rainbow. The primary rainbow's angular radius is about 42 degrees and its width is about 2 degrees. The secondary rainbow's angular radius is about 51 degrees and its width is about 3 degrees.

By definition, raindrops between the rainbows cannot send any light that contributes to either the primary or secondary rainbow. In other words, light that has been internally reflected once (the primary) or twice (the secondary) by water drops does not reach you from this part of the sky, so the sky looks comparatively dark there. This dark band is known as Alexander's dark band in honour of Alexander of Aphrodisias who discussed this phenomenon some 1800 years ago, and is most evident if the primary and secondary bows are bright.

The Antisolar Point

If we look at the ground on a sunny day, the shadow of our head marks the point called the antisolar point, 180 degrees away from the Sun. If the Sun is in the sky, the antisolar point is below the horizon. If the Sun has set, the antisolar point is above the horizon. What does this have to do with rainbows? The antisolar point tells us where we can expect a rainbow to form, since the coloured light from the raindrops exit those raindrops at specific angles that we can measure with respect to the antisolar point. In other words, any rainbow we see in the sky is due to the raindrops being 40.6 to 42 degrees from the antisolar point reflecting coloured light into our eyes.

What Causes the Rainbow's Circular Shape?

Many drops acting in concert cause the rainbow, and all of these must be at the same angle from the Sun (that is, the same angle from the antisolar point). Thus, at any instance only those drops that are on a 42 degrees circle centred about the antisolar point can send you the concentrated rainbow light. These drops may be at any distance, but they must be on the 42 degrees circle (Fig. 2.7).

Put another way, the rainbow is a mosaic of light sent to you by many raindrops as they fall through the surface of the imaginary cone whose tip is at your eye and whose radius is 42 degrees. If at any time, you see many sunlit drops in all directions from the antisolar point, the rainbow circle will be complete. For example, you could see the rainbow circle from atop a mountain,

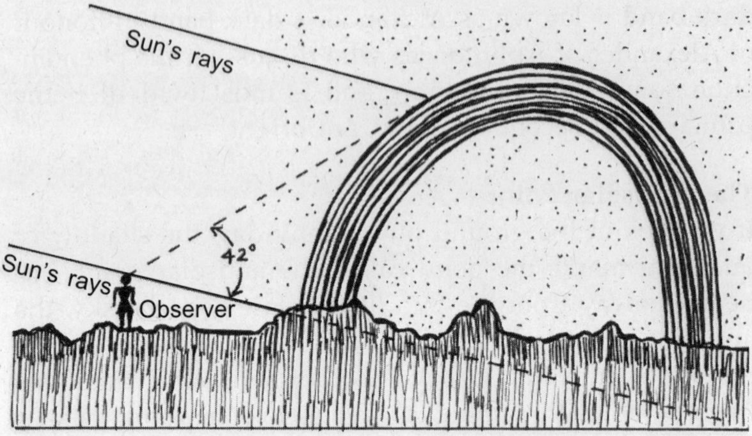

Fig. 2.7: How a rainbow is formed

a high hill or building, or an airplane in flight. Did you know that it is possible to see a rainbow in another shape other than the known arch? If you are ever in a plane and see a rainbow you may actually get the opportunity to see a round rainbow.

Green Flash

Sometimes you may spot a green light near the upper rim of the setting or rising Sun. This unusual phenomenon is known as the 'green flash'. Green flashes are usually only visible for a few seconds. However, Admiral Richard E Byrd saw one for a much longer time. While on an expedition to Antarctica in the winter of 1934-1935 he observed a green flash for 35 minutes.

An old Scottish legend says that if you have ever seen a green flash you will always be able to see into your own heart and also be able to read the thoughts of others.

Scientists say a green flash is caused by refraction differences between the wavelengths of light. Longer wavelengths of light are refracted less than the shorter ones. This means that near the horizon there is a difference in the bending of the red and blue wavelengths. The difference can be between 20 and 40 degrees. This difference makes you see two Sun discs that are partially covering each other. The violet-to-green arch is just a little bit higher than the red one. This means the violet and blue rays are disrupted coming through the atmosphere and we cannot see them. Instead, we see a 'green flash'.

Spotting a Comet

Have you ever wished on a shooting star or a comet in the sky? They say such wishes come true. So watch out for comets—Earth's celestial visitors that appear every five or six years. More spectacular visits happen about every 10 years. We can spot a comet in the sky with the unaided eye. The most remarkable sight is the comet's bright and shining tail.

A comet develops a tail as it nears the Sun. A comet's tail always points away from the Sun and can be up to 10 million kilometres long. It can leave behind trails of gas that can extend several hundred million kilometres further. Some comets actually have two tails—one of gas and another one inside made up of dust.

A comet resembles a fuzzy star that travels along a definite path through the Solar System. Icy chunks of water and dust form comets that originate in the outer Solar System. There are 'short-period comets', which take a short time to orbit the Sun and then there are the 'long-period comets', which take a long time to orbit the Sun.

Whenever the Earth passes through the tail of a comet, shooting stars dart across the sky. These are known as 'meteor showers'.

Eclipse–Darkness at Noon

The light of day suddenly begins to fade at noon. Looking up, you catch a glimpse of what looks like a disk of pure blackness sliding across the face of the Sun. Soon the blackness almost completely covers the Sun, and dusk falls over the land (Fig. 2.8). The air cools, the birds are silent and still. The scene is both frightening and beautiful. This is an eclipse—an overthrow of the natural order of things. The word eclipse comes from a Greek word meaning 'abandonment'. Quite literally, an eclipse was perceived as the Sun abandoning the Earth.

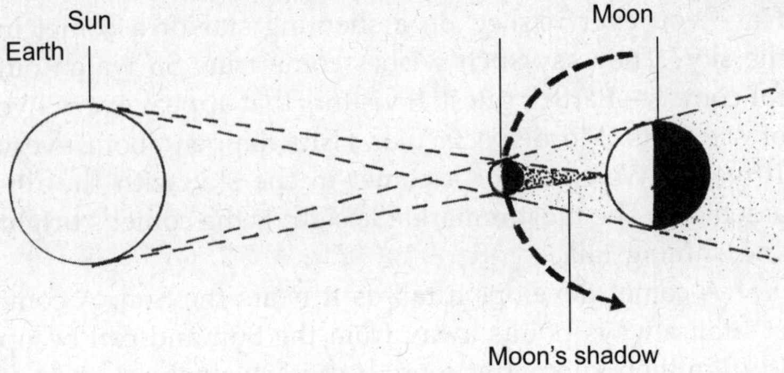

Fig. 2.8: Solar eclipse

A recurring and pervasive embodiment of the eclipse was a dragon, or a demon, which devoured the Sun. The ancient Chinese would produce great noise during an eclipse, banging on pots and drums to frighten away the dragon. The Incas, too, tried to intimidate the creatures that were apparently 'eating' the Sun. In India, even today people take a bath to purify themselves from the evil effects of an eclipse which they believe is caused when the demon *Rahu* gobbles up the Sun.

It does make sense that eclipses would be seen as bad omens. For most early cultures, the Sun was seen as a life-giver, something that was there every day, so anything that blotted out the Sun was terribly bad, the event itself filled with foreboding. Poet John Milton, in his epic poem *Paradise Lost*, captures the unease eclipses generated in early Europeans:

As when the Sun, new risen,
Looks through the horizontal misty air,
Shorn of his beams or from behind the Moon,
In dim eclipse, disastrous twilight sheds
On half the nations, and with fear of change
Perplexes monarchs.

As a scientific view of the universe grew in influence, the image of the eclipse began to shift. The new view reflected the scientific approach and rejected awe and terror, emotions that came to be seen as a sign of an undeveloped, uncivilized mind. The English poet William Wordsworth's *The Eclipse of the Sun*, 1820, describes the contrast between the attitudes:

High on her speculative tower
Stood Science waiting for the hour
When Sol was destined to endure
That darkening of his radiant face
Which Superstition strove to chase,
Erewhile, with rites impure.

The theme of science-versus-superstition appears in the world-renowned comic-book adventures of Tintin, the boy-reporter created by Hergé. In his well-illustrated comic book *Prisoners of the Sun*, Tintin and his companions are taken prisoners by a tribe of Incas. For their crime of accidentally entering the Temple of the Sun, Tintin and Captain Haddock are to be burned alive, their pyre lit by a huge magnifying glass focusing the rays of the Sun. The only concession the Incas give them is that they may choose the day and hour of their death. Tintin providentially enough, finds a scrap of newspaper in his pocket that predicts a total eclipse a few days hence. Choosing that day and time for their execution, Tintin then is able to halt the ceremony, shouting to the leader of the Incas "Stay, Husker! The Sun God will not hear your prayers! O magnificent Sun, if it is thy will that we should live, give us now a sign!" And the Sun, appearing to obey Tintin's will, begins to disappear behind the eclipse. The Incas are terrified, and rush about in chaos. The Inca leader frees Tintin and his friends, and they are accorded places of honour.

However, one must agree that it is quite natural to feel a sense of wonder tinged with awe when the Sun disappears.

What Is An Eclipse?

An eclipse occurs when the shadow of one object moves in front of another object or when one object moves in front of another to block its light. In simple words an eclipse is the darkening of a heavenly body. The Sun appears to turn dark as the moon passes between the Sun and the Earth causing a solar eclipse. A lunar eclipse occurs when the moon darkens as it passes between the Sun and the Earth. Sometimes planets or other heavenly bodies other than Earth and moon eclipse each other. For instance, Jupiter's moons cause shadows on the planet or Jupiter itself blocks Sunlight from its moons.

Solar Eclipse

A solar eclipse occurs when the moon passes between the Sun and the Earth. The Sun is blocked out by the moon and cannot be seen from areas on the Earth that lie in the moon's shadow. The shadow usually moves from west to east at a speed of about 3,200 kilometres per hour. People in the path of the shadow may see one of three kinds of eclipses: Partial eclipse, Annular eclipse, Hybrid eclipse and Total eclipse.

Partial – A partial eclipse occurs when the moon passes in front of the Sun but covers only a part of the sun.

Annular – If the moon is further away from the Sun than normal, then the moon's shadow will not be large enough to cover the Sun. So, the moon darkens only the middle of the Sun leaving a bright ring around the edges.

Hybrid – A hybrid eclipse is one, which appears as a total eclipse in some parts of the world, and as an annual eclipse in others.

Total – A total eclipse occurs when the moon covers the Sun completely.

Stages of a Total Solar Eclipse

1. A small bit of the Sun disappears.
2. The Sun gets completely covered
3. Totality—the period of time (usually few minutes)—when the Sun can be observed with the naked eye.
4. The Sun reappears from behind the moon – the diamond ring.
5. The Sun is completely restored.
6. Photosphere, the bright solar surface that we normally see and which is completely covered during the total eclipse can be seen again.

At the time of a total solar eclipse, a brilliant halo, the corona flashes into view around the darkened Sun (Fig. 2.9). The blue sky darkens and some bright stars or

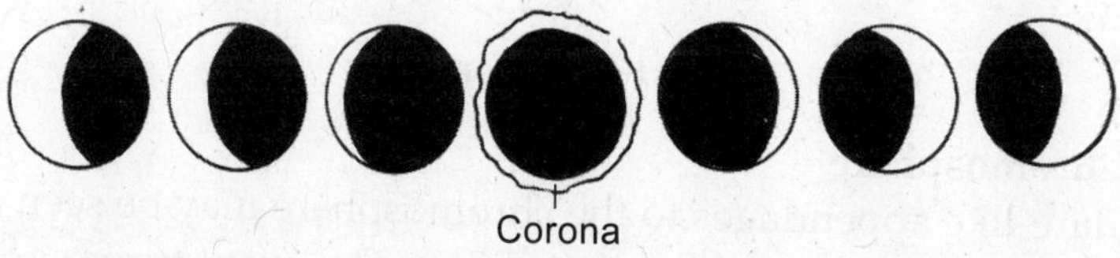

Fig. 2.9: Total solar eclipse

planets may even be visible. The Sun reappears as the moon moves off to the east. The Sun darkens totally for as long as 7 minutes and 40 seconds, but it averages about two and a half minutes. A total solar eclipse can be seen in those areas of the world, which lie in the path of totality, the path along which the moon's shadow passes across the Earth.

Bailey's Beads

These are a familiar feature of total eclipses. As the moon completely covers the Sun, the razor-thin solar crescent breaks up into a chain of beads that gradually wink out. Totality starts when the last one disappears.

The Diamond Ring

This is one of the most dramatic features of the eclipse. As the photosphere reappears, it briefly shines like a diamond. The rapidly fading corona appears like the rest of the ring. The picture as it appears in the sky is that of a spectacular cosmic diamond ring (Fig. 2.10).

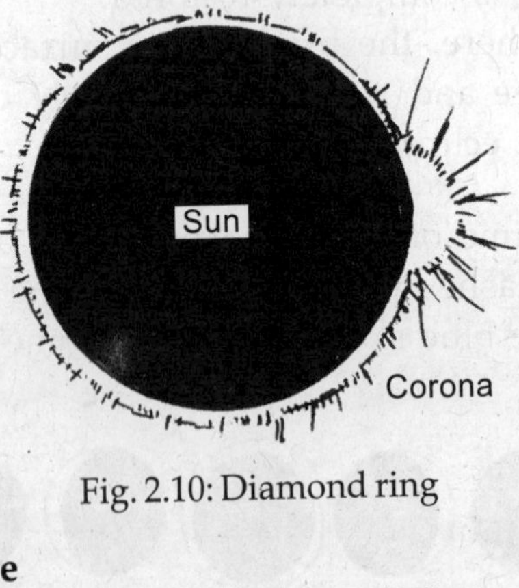

Fig. 2.10: Diamond ring

Chromosphere

Flare-like appendages to the chromosphere may be seen at any location round the eclipsed Sun. The event turns spectacular when the Sun is 'active'.

Spectroscopic studies of the Sun during solar eclipses indicate that a typical absorption spectrum is seen until the Moon gets to the bottom of the chromosphere. Then the spectrum reverses itself to an emission spectrum. This has led to the bottom-most portion of the chromosphere to be called the reversing layer. The corona or the 'crown' of the Sun extends outward from the chromosphere to a distance of several million kilometers where it gradually becomes the solar wind. The corona does emit light of its own, but this light is only as bright as the light of a Full Moon (about one-millionth as bright as the photosphere), so it is normally invisible. But when the Moon blocks out the much brighter photosphere during an eclipse, the corona shines forth with its eerie opalescence.

How Often Do Eclipses Occur?

A total solar eclipse is one of the nature's most impressive sights. Total eclipses happen infrequently. The orbits of both, the Earth around the Sun and the orbit of the moon around the Earth are elliptical. Also, these orbits do not lie parallel to each other in the same plane. As the Earth orbits the Sun, taking one year to complete one circle, it appears to us on Earth that the Sun moves around our sky once against the background of stars. Let's say, if you walk around a campfire (the Sun) looking at your friends on the other side (the stars), to you it would look like the campfire moves past your friends. Likewise, from Earth, it looks like the Sun moves against the background of stars, making one round of the sky in one year.

If the Sun could draw a line as it moved against the stars, we would see an oval called the ellipse. If we could also ask the moon to draw a line in the sky as it orbited the Earth, we would notice that the two lines would be close to each other, except that the moon's path is tilted about 5 degrees to the path of the Sun. This is precisely why the moon does not eclipse the Sun every month. Most of the time, the moon passes over or under the Sun. An eclipse can happen only when both the Sun and the moon arrive near one of the crossing points (these are called nodes). There are two nodes on opposite sides of the sky, one where the moon crosses from south to north, and one where the moon passes from north to south. Since there are two crossing points in the sky, eclipses happen during two 'eclipse seasons' separated by about six months.

The type of eclipse that does occur depends on several things. First, if the eclipse happens when the Sun is further from the node, it is more likely that the eclipse will be a partial one. In this type of eclipse, the dark umbra passes above the North Pole or below the South Pole, never touching the Earth. All we ever see is part of the Sun covered.

To See a Solar Eclipse

Solar eclipses happen about twice a year, but not always in the same part of the world. To see a total solar eclipse, you have to be in just the right spot. When you are standing on the Earth, looking up at the Sun and the moon, you must be in a very limited zone to see the moon cover the entire face of the Sun. If you were to move a little north, the Sun would peek out over the top of the moon; a little south, and the Sun, shines past the southern limb of the moon. The match is so good that the 'path of totality' is never more than 267 km in diameter, and is usually less. This means that very few people have seen a total eclipse because the shadow only covers a very small area on the Earth.

Dangers of Viewing a Solar Eclipse With the Naked Eye

Never view a solar eclipse with the naked eye. Radiation from the corona can damage the eyes. If you look at the Sun, your eye-lens will concentrate on the Sun's light and focus it to a very small spot on the back of your retina. This can cause permanent eye damage or blindness. Additionally, there are no pain sensors there, so you would not even know it's happening! The use of any optical device, such as binoculars, camera, telescope, smoked glass, darkened film or sunglasses may not eliminate the danger of watching a solar eclipse. During totality, it is safe to view the eclipsed Sun. But you should make sure to turn away before the Sun reappears. Even glancing at the 'diamond ring' is dangerous. The eclipse can be viewed through a pair of special filtering glasses or indirectly with a pinhole projector (Fig. 2.11).

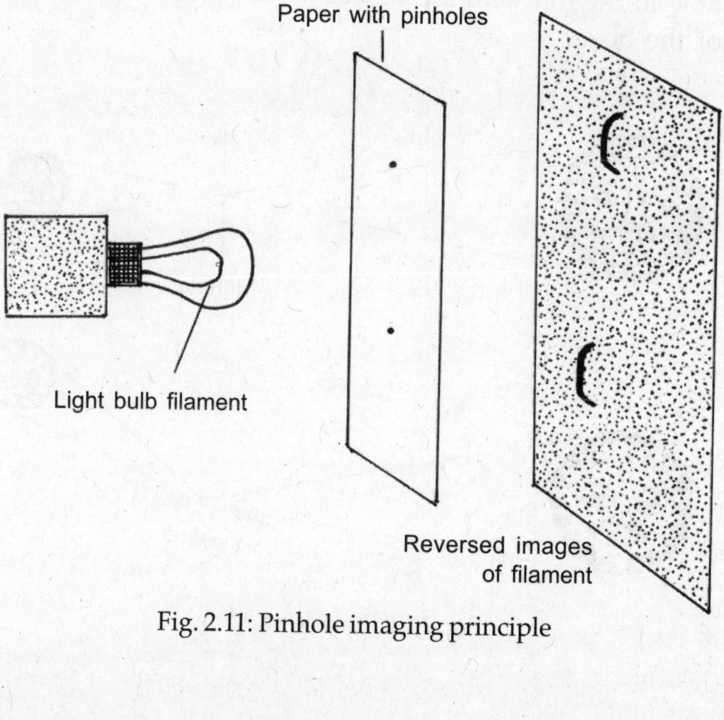

Fig. 2.11: Pinhole imaging principle

Make Your Own Pinhole Viewer

- Take two sheets of cardboard, and make a pinhole in one.
- Stand with your back to the Sun, and hold both sheets up, with the hole facing towards the eclipse.
- Experiment by holding the sheets at different distances apart until you see an image of the eclipse appearing on the second sheet.

Make Your Own Pinhole Box to View the Sun

Material required – A long box, at least 180 cm, a piece of aluminium foil, a pin, and a sheet of white paper. The longer the box, the bigger the pinhole image will be. The size of the image will be approximately about 1/100th the length of the box. A long tube will also do.

Method – Cut a hole in the centre of one end of the box. Tape a piece of foil over the hole and poke a small hole in the foil with a pin. Cut a viewing hole in the side of the box.

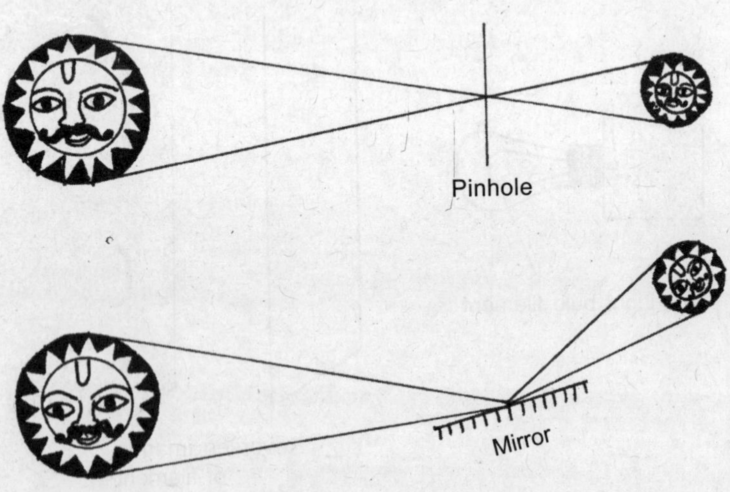

Fig. 2.12: Pinhole imaging

Now put a piece of white paper inside the end of the box near the viewing portal. Point the end of the box with the pinhole at the Sun so that you see a round image on the paper at the other end. If you are having trouble pointing, look at the shadow of the box on the ground. Move the box so that the shadow looks like the end of the box (so the sides of the box are not casting a shadow). The round spot of light you see on the paper is a pinhole image of the Sun. Do not look through the pinhole at the Sun! Look only at the image on the paper (Fig. 2.11).

Studying Eclipses

Astronomers learn a lot by studying eclipses. The famous physicist Albert Einstein theorised that light from stars and beyond the Sun bends slightly as it passes the Sun. This starlight can be photographed during a total solar eclipse. In 1675, the Danish astronomer Olaus Roemer calculated the approximate speed of light by studying Jupiter's eclipses of its moons. Scientists study eclipses by making artificial eclipses with an instrument called a 'coronograph'.

Artificial Eclipse

Space-based solar observatories, such as the Solar Heliospheric Observatory (SOHO), employ a kind of artificial eclipse to routinely view the Sun's corona. The eclipse is produced from a solid disk mounted in the observatory's detector. This disk blocks light from the Sun's surface, just as the moon blocks our view of the Sun during a solar eclipse. But the detectors on space observatories are not able to view the part of the Sun's atmosphere just above the surface, the layer where violent solar storms gather their energy. Contained by the Sun's powerful magnetic fields, these storms brew until they burst through in an exploding mass of charged gas that can rush toward the Earth at speeds of up to 2000 km per second.

These are violent solar explosions called Coronal Mass Ejections, or CMEs. When these CMEs are directed toward the earth, turbulent shock waves of charged gas and their accompanying magnetic fields can interfere with radio, television, and telephone signals, damage satellites and disrupt satellite communications, and build up voltages in electric power lines. High-energy particles from these storms also contribute to the beautiful polar light shows known as *aurora borealis* in the northern hemisphere and *aurora australis* in the southern hemisphere. At times the most violent solar storms threaten the safety of space-walking astronauts and generate huge geomagnetic storms on Earth. In 1977, a CME knocked out a communications satellite and silenced pagers all over the world. In 1989, a direct hit by a CME knocked out power to Quebec (Canada) and produced significant damage in satellites orbiting the Earth.

Tides

We think of tides in association with the sea. Tide is the rise and fall of large expanses of water according to a definite time schedule. We love to swim when the tide is 'in' and then look for seashells when the tide is 'out'. Tides are important for ships that leave or enter the harbour and they signal the start and the end of the working day for fishermen who go fishing at scheduled times.

We realise that the moon causes the tides but not everyone realises that the Sun also plays a role in this phenomenon. As the Earth rotates on its axis, first one place and then another falls directly beneath the Sun or the moon. The moon and Sun's gravitational attraction at these points is so great that it actually lifts the ocean and bulges the crust of the Earth. Troughs develop where the gravitational attraction of the Sun or moon is the weakest.

Forces of the Sun and moon bring about tides in all bodies of water, large or small. Approximately twice a day, the level of the sea rises; when the level is highest, it is referred to as high tide. When the sea level is lowest, midway between high tides, it is referred to as low tide. The interval between two moonrises is about 24 hours 50 minutes. The time between the two rising moons is determined by two motions, the rotation of the Earth on its axis and the revolution of the moon around the Earth.

Since the moon is four hundred times closer to the Earth than the Sun is, it has more influence on tides than does the Sun. The Roman naturalist Pliny before the year AD 100 wrote about the moon's influence on the tides. The English scientist Sir Isaac Newton was the first person to explain tides scientifically. His explanation of the tides (and many other phenomena) was published in 1686, in the second volume of the 'Principia'. He worked out the physical laws of tides following the discovery of the law of gravitation.

Causes of Sea Tides

The pull of the moon on the Earth causes the tides in the sea. The moon's gravity pulls up the water directly below the moon, forming a high tide there. High tide also occurs on the other side of the Earth because the moon pulls the solid Earth away from the water. As the Earth turns, high tide occurs at each place on the sea twice a day.

What does the pull of the moon do to the oceans? It creates two types of tides, high and low. As the moon rotates around Earth, tidal bulges occur. The bulge is really a large wave beneath the moon that moves across the Earth. On the opposite side of Earth, there is a second bulge. These bulges are high tides. Between each high tide, there is a low tide, which is called ebbing. Usually two high and two low tides occur every 24 hours and 50 minutes, because that is how long it takes the moon to rotate around Earth.

Seawater does not follow the moon or go to the middle of the ocean, it stays in the same place. You can carry out an experiment to demonstrate what happens. Place a coin on a table. Now pass a magnet over this coin. The magnet is not close enough to actually pick up the coin, but instead makes an edge of the coin rise a few millimeters. This is similar to what the moon does to the ocean. Like the coin, the ocean rises periodically because it is attracted to a force above it (the moon). Neither the water in the ocean or the coin move or follow the above force, they just react to it.

The difference between high water and low water is called the range of the tides. The characteristics of the seacoast makes a difference in the range of the tide. In funnel-shaped estuaries and bays, the range may be very high. In the Bay of Fundy, Nova Scotia, Canada, for example, the difference between high and low tide is sometimes more than 15 metres. The shape, size and depth of seas or oceans also make a difference in the way the tide acts. For instance, the Atlantic Ocean has tides that flow and ebb regularly twice a day. Some Pacific Islands have mixed tides such as two high tides daily with only a little ebb between, and then a very low tide. In places like Alaska, or along the Gulf of Mexico, there is only a daily tide, a low and a high tide each day. The Mediterranean Sea, on the other hand has little tidal range.

Spring Tides

Spring tides have nothing to do with spring season. They

are especially strong tides which occur when the Earth, the Sun, and the moon are in a line. The gravitational forces of the moon and the Sun both contribute to the tides. When the moon is full or new, the gravitational pull of the moon and Sun are combined. At these times, the high tides are higher than the normal and the low tides are very low. These are known as spring tides. Spring tides occur twice a month, near the times of full and new moons. The moon lies either between the Earth and the Sun or on the opposite side of the Earth from the Sun (Fig. 2.13).

The proxigean Spring tide is a rare, unusually high tide. This very high tide occurs when the moon is both

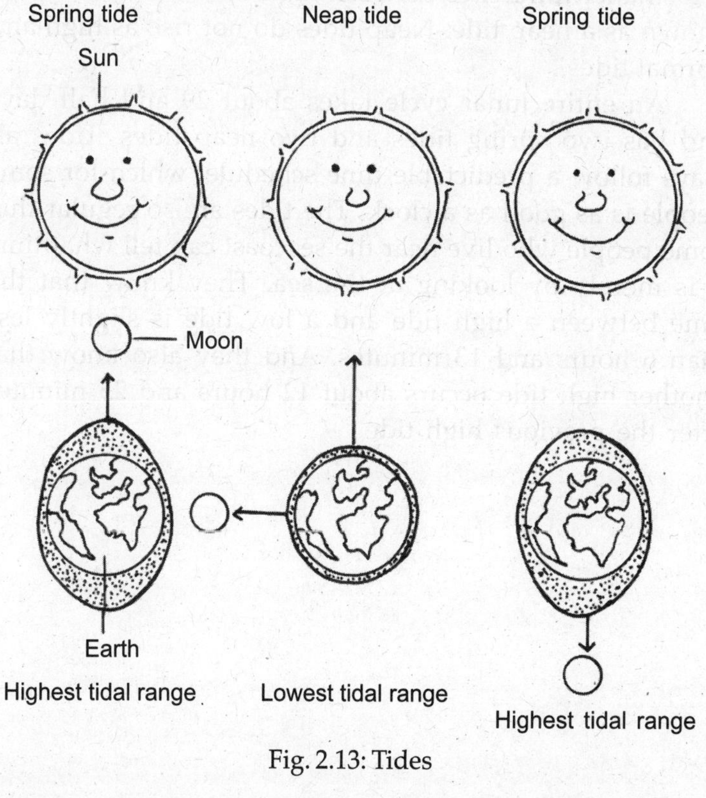

Fig. 2.13: Tides

unusually close to the Earth (at its closest perigee, called the proxigee) and in the New Moon phase (when the Moon is between the Sun and the Earth). The proxigean spring tide occurs at most once every 1.5 years.

Neap Tides

Neap tides are especially weak tides. They occur when the gravitational forces of the Moon and the Sun are perpendicular to one another (with respect to the Earth). During the moon's quarter phases, the first and the last quarters, when the moon is either on the side of the Earth or on the opposite side of the Earth, the Sun and moon work at right angles, causing the bulges to cancel each other. The result is a smaller difference between high and low tides and is known as a neap tide. Neap tides do not rise as high and normal tides.

An entire lunar cycle takes about 29 and half days and has two spring tides and two neap tides. Tides always follow a predictable time schedule, which for some people is as good as a clock. The tides are so regular that some people who live near the seacoast can tell what time it is merely by looking at the sea. They know that the time between a high tide and a low tide is slightly less than 6 hours and 13 minutes. And they also know that another high tide occurs about 12 hours and 25 minutes after the previous high tide.

Plate 1: Diamond Ring (*Courtesy*: NASA)

Plate 2: The Planet Earth (*Courtesy*: NASA)

Plate 3: Planets of the solar system (*Courtesy*: NASA)

Plate 4: Corona of the Sun (*Courtesy*: NASA)

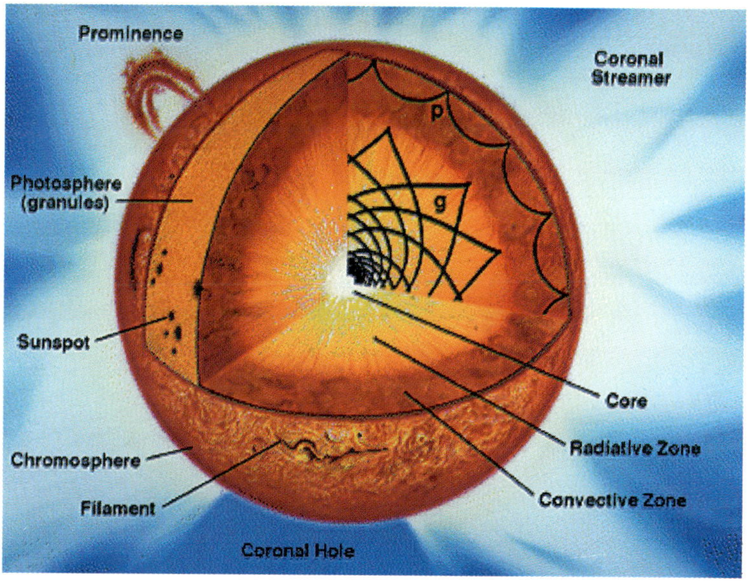

Plate 5: The interior of the Sun (*Courtesy*: NASA)

Plate 6: Sun Flares (*Courtesy*: NASA)

Plate 7: Sun Umbra (*Courtesy*: NASA)

Plate 8: Total solar eclipse of 21 June 2001 (*Courtesy*: NASA)

3

PLANTS AND THE SUN

Have you any plants growing in flowerpots at your home? Do you look after them? What do plants really need? All plants need is Sunlight and water (Fig. 3.1). Without any light at all, most plants would not do very well, and many would die. Both indoor and outdoor plants need Sunlight, water, air, minerals and nutrients to grow strong and healthy. Why do the shoots grow upwards? They are looking for the Sunlight that they need for photosynthesis, which is the way they make their food. The shoots try to get as much Sunlight as possible. The beautiful colour of plants and flowers depend on the amount of Sunshine they are exposed to. Sunlight provides the energy that plants need to make the sugars and starches

Fig. 3.1: Plants need sunlight to grow

they use as food. Plants grow towards Sunlight and the amount of daylight received by a plant affects its growth.

The Sun effects plants and gardens and this is particularly noticeable when there is either too little or too much of it. It can be the cause of feast or famine for many a farmer.

Photosynthesis

The colour of grass is green. Most garden plants are green. The reason for their green colour is simple. Most plants are autotrophs, which produce their own food from energy they get from the Sun. They are able to make their own food using the green coloured matter called chlorophyll and a process called photosynthesis. But they need sunlight, carbon dioxide and water in addition to cells containing chloroplasts to make food for them. The word photosynthesis means 'putting together with light'.

Plants get carbon dioxide from the air. The air enters the leaves through tiny holes called stomata. Plants get the water they require through their root systems in the soil.

When energy from Sunlight is available, the process of photosynthesis takes place. Photosynthesis involves only four major ingredients: water, carbon dioxide, chlorophyll, and Sunlight. Leaves are like small factories that produce food for the plant. Photosynthesis takes place in small bodies called chloroplasts within the cells of plant leaves. Inside these cells are tiny lens-shaped structures called the chloroplasts, which are only 0.004 millimetres across. Each lens-shaped chloroplast consists of green granas within a colourless stroma. It is inside the grana that the chlorophyll captures the Sun's light energy. These choloroplasts contain chlorophyll, which is bright green in colour, and which absorbs sunlight.

When the light strikes a molecule of chlorophyll, its

electrons absorb energy. This excess energy excites the chlorophyll's electrons so much that they bounce from one chlorophyll molecule to another. As the excited electrons bounce around, they split apart molecules of water, producing both hydrogen and oxygen.

During the process of photosynthesis, the energy from the Sun is turned into chemical energy. Energy from the Sun splits water molecules into hydrogen and oxygen. The oxygen is released into the air, and the hydrogen combines with carbon dioxide to form the basic building blocks, which plants use to manufacture food. The hydrogen joins with carbon from the carbon dioxide to produce sugar. The sugar helps a plant to make the fat, protein, starch, vitamins and other materials that it needs to survive (Fig. 3.2). The process releases oxygen into the air. People and animals need the oxygen to breathe.

$$\underset{\substack{\text{Carbon}\\\text{Dioxide}}}{CO_2} + \underset{\text{Water}}{H_2O} + \text{Chlorophyll} + \text{Sunlight} = \underset{\text{Oxygen}}{O_2} + \text{Sugar}$$

The basic food, or sugar, that plants make during this process is called a carbohydrate. The plant uses some of these sugars for its own needs and then stores some of the sugars. One very common carbohydrate is a sugar called glucose. When glucose is combined with nitrogen it forms amino acids, proteins or nucleic acid. Nitrogen is found dissolved in the water absorbed by the roots. Glucose can also be converted to cellulose, which is used to build plant tissues. When converted to starch, the glucose is stored in the leaves, stems and roots for later use. For instance the potato plant stores its sugar, or carbohydrate, in underground tubers such as potatoes.

On overcast days or during winter when the angle of the Sun decreases, photosynthesis stops because there is not enough energy to split the water molecules apart.

Fig. 3.2: Process of photosynthesis

The process of photosynthesis also breaks down at higher temperatures because the enzymes of many plants fail to function properly at high temperatures. This means many plants stop producing food on hot summer afternoons.

Reaching for the Light—Phototropism

Sunlight, climate and soil condition—these are important environmental factors that shape the growth of a plant.

Heredity determines a plant's characteristics such as size, shape and colour of a flower. Sunlight is very necessary for some species of plants without which they will die. There are other plants that thrive in the shade. However, shade-loving or not, all plants need the Sun.

A plant may display a bending movement called as tropism. Tropism comes from the Greek word *trope*, meaning, 'to turn'. In tropism, an outside stimulus or force causes a plant to bend toward or away from the stimulus. There are varied tropisms, for instance, *phototropism*—caused by light, *geotropism*—caused by gravity and *hydrotropism*—caused by nutrition. Tropisms may be positive (bending toward the stimulus) or negative (bending away from the stimulus).

Phototropism is commonly seen in plants, as they tend to grow toward the Sunlight. A plant placed in a window-sill exhibits positive phototropism when its stems and leaves grow toward the source of light. Roots, on the other hand show negative phototropism growing away from light. The leaves and stems of plants always grow in the direction of a light source. So even if it has to twist in some odd directions, the plant will keep reaching for the light! They will do everything they can to get as much of it as possible (Fig. 3.3). Since most plants cannot really move, they grow towards the sun instead. When an old tree dies, it

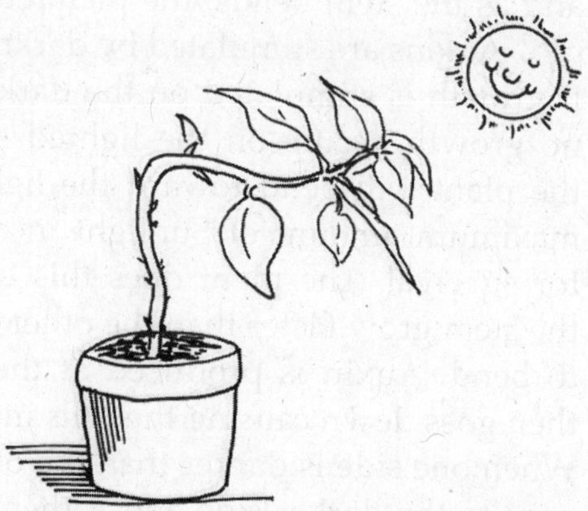

Fig. 3.3: Plant bends towards sunlight

leaves an open space where more light reaches down to the forest floor. Then small plants start growing, reaching up to get as much of that light as they can.

The best example of phototropism is shown by the sunflower. The yellow sunflower head turns and faces toward the Sun throughout the day. This is because the side of the Sunflower's stalk that is away from the Sun grows faster than the side facing the Sun. The stalks actually seem to be following the Sun. The movement of the Sunflower stalk allows the leaves to receive the Sun's light most effectively.

Studies of tropism led to the discovery of auxins, the plant hormones that control growth. Phototropism is the result of auxin action. Auxin literally means to increase. It is present in all green plants. It has the peculiar characteristic of always moving away from the lighted side to cells on the shaded side of the plant. High concentrations of auxin cause stem cells to grow more rapidly and to elongate. Its absence causes cells to slow their growth and to shorten. This causes a stem to grow unevenly. The side away from the light grows more rapidly than the side toward the light. As a result, the stem begins to bend, and as the stem bends, the plant curves toward the light.

Auxins are stimulated by darkness and cause growth. If growth is stimulated on the dark side of the plant and no growth occurs on the lighted side of the plant then the plant will bend toward the light and try and gather maximum amount of Sunlight in order to produce food for survival. The plant does this by making one side of the stem grow faster than the other side, causing the plant to bend. Auxin is produced at the tip of the plant and then goes down causing the cells in the stem to get longer. When one side is darker than the other, more of the auxin goes to the darker side and it then grows faster, bending the stem toward the light.

Plants Know When to Flower—Photoperiodism

Have you ever wondered why some plants flower in the summer season while others do it in winter? Temperature too plays a part. Bulbs are an example of this—they produce flowers after a period of cold climate. Plants are unable to know what time of the year it is at any given time but they can work it out by progressively measuring the length of day-night time. This is called photoperiodism. The lengths of periods of light and dark that plants receive affect the growth of plants. Some plants need long periods of daylight while others bloom only when the dark period is long.

The Sun regulates the schedule of plants. For instance, in the Northern Hemisphere, the length of time and light constantly varies. In summer the days are long and it remains well-lighted, sometimes even until 3 a.m. in the morning. In winter it turns dark sometimes at 4 p.m. in the afternoon! The only times the length of time of the daylight and the darkness are equal are on March 21st and September 23rd. On every other day of the year, the lengths of the day and the night are unequal. Each day the amount of sunlight plants get varies. After each change the plants are triggered to do certain things. These lengthening or shortening periods of daylight serve as alarm clocks which signal plants when it is time to flower, drop their leaves, or prepare for winter.

Three aspects of light—quality, quantity and duration affect plant growth. Amongst these, quality of light is important for plants. We know that sunlight consists of a wide range of wavelengths, some of which we cannot see. Artificial light can come close to duplicating the full range of wavelengths found in sunlight. One warm white and one cool fluorescent bulb used together come close to producing the full spectrum of light wavelengths for indoor growing. Special 'grow lights' also produce this quality of light.

The amount of light that plants receive depends upon seasons. Plants receive maximum quantity of sunlight in summer and minimum in winter because of the angle at which the wavelengths of light hit the plant surface. Duration is the time that a plant is exposed to light. This is critical to growth and flowering. Outdoors, the duration of light varies with the season. Summer days have a longer duration of light than winter days. Some plants flower in response to the length of darkness they experience. The time that a plant is exposed to light is critical for growth and flowering.

Flowering plants can be of three categories. The long-day plants need a daylight period of over 12 hours. The short-day plants need a daylight period of less than 12 hours and the day-neutral plants are not fussy about the day length. Some plants, such as the lettuce and spinach, bloom only when the photoperiod is long. Chrysanthemums, asters and poinsettias are short-day plants while the length of the photoperiod does not affect plants such as tomatoes and marigolds. The latter are the day-neutral plants that do not depend on day length to form flowers, as long as there is enough light for normal growth.

The reason the plants are triggered by the changing amounts of light is because of a substance called 'photochrome'. Photochrome inside the cells of most plants measures the length of the day by absorbing light. It causes short-day plants such as chrysanthemums, ragweeds, goldenrods, soyabeans, and poinsettias to flower during the spring and autumn. Long-day plants such as hollyhocks, clover, irises, radishes and beets bloom in the late spring and early summer when the days are longer. Photochrome also lets a plant know when winter is coming. As the number of hours of daylight decreases in the autumn, photochrome stimulates plants to form winter-resistant buds.

Photochrome also stimulates leaf stems to seal off the flow of water and nutrients to a plant's leaves. When the length of day begins to shorten, twigs form a cork-like plug called an 'abscission zone', at the base of each leaf. As the plug thickens, it literally strangles the leaf to death. When the leaf dies, its chlorophyll starts to fade and it loses its green colour. This fading unmasks the plants true colours causing the plant to turn a bright red eventually fading into orange and then a yellowish colour. This reveals rich colours, which have been masked all summer long by the abundance of chlorophyll. Eventually the leaf breaks off at its base and falls to the ground.

4

ANIMALS AND THE SUN

"The Sun's over there so my home must be this way", thinks a homing pigeon. Long, long ago before there were telephones or cell phones, people used birds to carry messages. Homing pigeons have an uncanny sense of direction. They depend on the position of the Sun and their own body clock to find their way back to their starting places. Birds can detect minute motions of the Sun that the human eye cannot. A bird's internal clock allows it to adjust its flight as Sun's position changes. At night they use patterns of stars or navigate by North Star. Not only birds but other animals too rely on the Sun for food, navigation, migration, etc.

Light and heat from the Sun gives all living organisms energy for growth, movement, reproduction and feeding. Plants use Sunlight directly in photosynthesis. Animals, on the other hand, cannot directly use the energy from the Sun; they have to rely on the photosynthetic organisms for this. The colour of butterflies, birds and animals are also determined by light, as is their complete development. Development of a tadpole into a frog is a good example. The process of metamorphosis is arrested if it is deprived of sunlight. Sunlight also enables the body of the animal to assimilate calcium. Birds raised in the Sunlight produce harder and thicker shells than those not so exposed.

Many animals respond to changes in the amount of

light available. The annual migrations, hibernation, metabolism, mating seasons, and molting times are all triggered by Sunlight. Birds more than any other species are profoundly affected by light because they rely on the sensory input from their eyes. They obtain vitamin D_3 from both diet and the synthesis in their skin. The synthesis of vitamin D_3 in the skin requires ultraviolet (UV) light, which is procured from the Sun's rays. Sunlight prevents infection. It stimulates male songbirds to sing. It also has a profound effect on avian reproduction behavior. Mating too is affected by light.

Seasonal changes greatly influence the plants and animals living on the Earth. Snowshoe rabbits, for example, completely change their appearance in autumn. They gradually begin to lose their brown summer fur and white winter fur grows in its place. This camouflage makes the snowshoe rabbit almost invisible against the background of the white snow. As the days grow longer in the spring, snowshoe rabbits shed their white winter coats. The white insulating hair of winter falls out and is replaced with darker summer hair. This summer hair is lighter in weight so it is cooler; and it allows the rabbit to blend in with the colours of summer.

Let's see how the Sun affects a lizard's day in a tropical region! (Fig. 4.1)

0100 hours At night when the air is cold, the body of the lizard cools down. The lizard sleeps in its hole.

0800 hours The Sun rises and the air warms up during the morning. Eventually, the sleepy lizard crawls out of its hole into the Sunlight. The lizard finds a piece of rock or sandy ground that is exposed to the Sun and lies there, basking in the Sun. Thus it warms its body up above the air temperature.

1000 hours	The lizard is active now, hunting for food and defending its territory.
1200 hours	The temperature in full Sunlight can reach 40°C or 50°C in tropical regions. This could be dangerous for the lizard; it could become too hot. The lizard retreats to its hole to find some shade.
1700 hours	The Sun's heat fades so the lizard can come out of its hole to hunt again.
2100 hours	The Sun goes down and night falls. Gradually the air becomes colder and the lizard returns to its hole to sleep.

Animals Use Sun As a Compass

Birds use the sky as a means of extracting compass information and they are capable of compensating for the movement of the Sun across the sky. Bees use polarised light as a celestial compass. Like the bees, the birds indicate the direction that they wish to fly by a well-known phenomenon called the *zugenruhe* or nighttime restlessness. *Zugenruhe* refers to the sharp jump in nighttime activity that occurs at the onset of the migratory response. Researchers suggest that birds take a bearing on the Sun at sunset and then use dead reckoning to fix on a correct direction until sunrise at which point a new bearing can be used.

Fig. 4.1: The garden lizard

Monarch butterflies (*Danaus plexippus*) from the eastern North American population use a time-compensated Sun compass during their migratory flight. They make remarkably long migratory journeys in the autumn, some extending more than 3,500 km from eastern USA and Canada to over-wintering grounds in the neo-volcanic belt in Central Mexico.

Caribou, salmon, and even turtles—all respond to the changing length of days throughout each year. Even not so gradual changes in light affect animals. Pet cockatoos are known to be quiet and go to sleep when their cage is covered with a sheet, even in broad daylight.

Navigation and Migration

Many birds, fishes, insects and mammals move from one place to another for better living conditions. Migrations take place in land, water and air for short distances as well as long distances.

Seasonal migrations are characteristic of many animals, such as fish, amphibians, reptiles, spiders, crustaceans, mammals and birds. Birds are known to

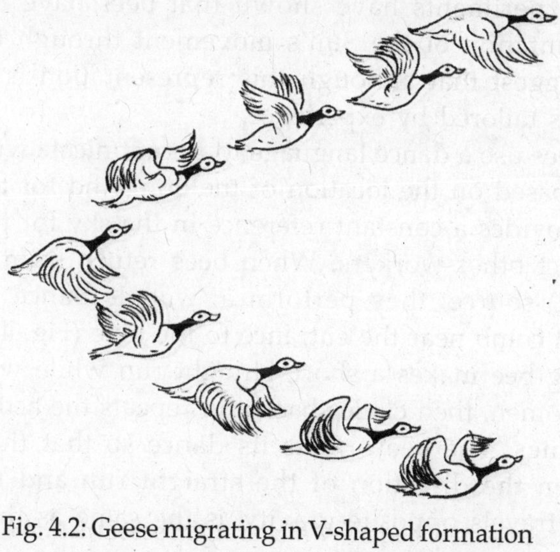

Fig. 4.2: Geese migrating in V-shaped formation

migrate to considerable distances, across continents and oceans, regularly and accurately (Fig. 4.2). They migrate from one part of the world to another, once in autumn from their breeding ground and again in spring from their wintering quarters.

Changing amounts of daylight prompt geese to fly south in autumn and north in the spring. The longest migrations are those by the Arctic Terns that leave their Arctic breeding ground in the autumn and fly to the Antarctic and then fly back to the Arctic in the spring, covering almost 35,000 km in a year. Many birds fly at night for a long distance and then rest for a few days. However, birds like geese fly non-stop from Canada to the southern states of the USA. Thus, the Sun serves as a guide for many migrating species.

Finding Their Way

How do bees find their way back to the hive once they are out on a nectar-hunting spree? Honeybees use the Sun as a reference point in navigation and communication. Experiments have shown that bees have internal representation of the Sun's movement through the sky and suggest that although this representation is innate, but it is tailored by experience.

Bees use a dance language to communicate with each other, based on the location of the Sun. And for this, the Sun provides a constant reference in the sky for the bees to direct other workers. When bees return from a food (nectar) source, they perform a 'waggle dance' on the vertical comb near the entrance to the hive (Fig. 4.3). The dancing bee makes a short, straight run while waggling its abdomen, then circles back and repeats the action several times. The bee orients its dance so that the angle between the direction of the straight run and the ray, which travels opposite gravity, is the same as the angle

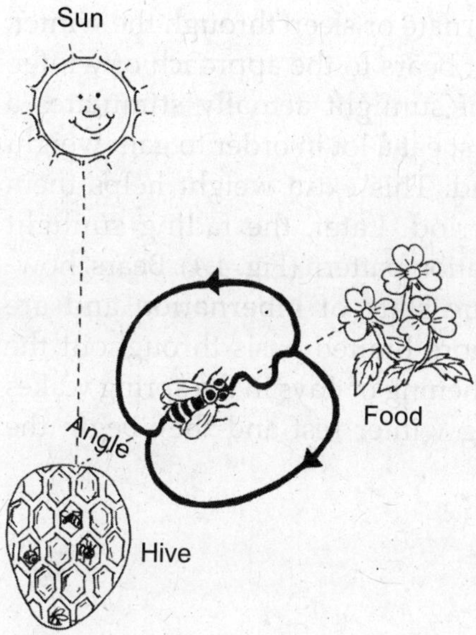

Fig. 4.3: The bee 'waggle dance'

between the food source and the position of the Sun. In this scheme, dancing straight up means 'fly toward the Sun', and dancing straight down means 'fly away from the Sun'. Given this angle, other bees can orient themselves to the Sun and locate the food source.

Hibernations

Animals face frequent danger from bad weather including cold, heat or drought. In places, which turn very cold due to minimal Sun reaching them, many animals such as bears, bats and frogs, burrowing

Fig. 4.4: Hibernating bear

animals, insects, etc. hibernate or sleep through the winter. The shortening days alert bears to the approach of winter. The changing amount of sunlight actually stimulates a bear's appetite. Bears must eat a lot in order to gain weight for the long winter ahead. This extra weight helps them store energy for this period. Later, the fading sunlight triggers the bear's hibernation pattern (Fig. 4.4). Bears, however, do not enter a true state of hibernation and are somewhat active on a more limited basis throughout the winter months. The lengthening of days in the spring wakes the bears from their long winter rest and they begin the cycle again.

5

ENERGY OF THE SUN

Let the Sun shine! Not only the Sun but also all stars produce huge amounts of energy. But the Sun is the powerhouse. In one second it produces 200,000 million times more energy than all the power stations on the Earth. There is immense solar power that can be harnessed and tapped. The Sun plays an important role in renewable energy. Through the Sun's heat, we can directly heat buildings and water. Indirectly, the Sun's heat creates the wind energy that can be harnessed through wind turbines. The Sun also warms the surface of the ocean more than the ocean depths. This temperature difference can be used as a source of energy. Sunlight can be converted directly into electricity through the use of solar panels.

Society has always relied on solar energy. Today non-renewable energy concentrates such as fossil fuels are being used up at an alarming rate. Energy conservation is one method of slowing the dwindling consumption of our fossil fuels.

Alternative energies such as wind, geothermal, water and especially solar energy show promise for our increasing future energy needs. Nuclear power is dangerous, and yet the Sun is a hydrogen fusion power plant, which is perhaps the best nuclear power plant we could hope to benefit from. It has a life expectancy of billions of years as opposed to the 20-year breakdown of a conventional U235 or Plutonium based plant. In the

21st century, the Sun's energy has become an increasingly attractive source of direct power to meet human needs.

All Energy Comes From the Sun

Coal, oil and natural gas are the fossil fuels. These are the remains of the plants and animals that died a long time ago. When these plants and animals were alive, they converted energy from the Sun into chemical energy. Now when we burn fossil fuels, we turn this chemical energy back into heat energy and light energy. In simple terms, we are releasing from these fuels the energy that first came to the Earth from the Sun millions of years ago.

The energy contained in the above-mentioned fossil fuels came from the Sun's energy, and it was stored in plants millions of years ago. But once used, these fossil fuels cannot be regenerated on a human time scale. The remaining fossil fuel of the Earth reserves can probably provide us with energy for another 100 to 500 years, but this is an insignificant amount of time in terms of the whole past history of human civilization and its future. The flow of renewable solar energies on Earth is essentially equal to the flow of energy due to solar radiation.

Solar Energy

Imagine that it's a hot summer day. You spill a scoop of ice cream on the sidewalk, and it melts. Why? Well, it's the Sun that caused the ice cream to melt. Energy given off by the Sun consists of light, heat and electromagnetic radiation. This is solar energy that is produced by nuclear reactions taking place inside the Sun. We use only a fraction of the solar energy that reaches the Earth. Much of the Sun's energy therefore remains to be captured and put to use where and when needed. The major problem with solar energy, however, is the intermittent and variable manner in which it arrives at the Earth's surface and the large area required to collect it at a useful rate.

Let us see how solar energy affects us. Life on earth depends upon the Sun's energy for heat and light and food. Plants, we know use solar energy to produce food during photosynthesis. This energy is recoverable through burning of wood and fossil fuels such as coal, petroleum, and natural gas. Plants are eaten by animals, which, in turn, are eaten by other animals. Solar energy provides warmth for plants and animals. Solar energy heats homes and greenhouses, produces wind power and generates waterpower through evaporation and rainfall.

Energy from the Sun sets the Earth's weather in motion. Heat from the Sun causes water on the Earth's surface to evaporate and form clouds that eventually provide fresh rainwater. Winds occur because the Sun's rays are more direct and thus stronger at the equator than they are at poles. The strong rays in tropical regions warm the air there, causing it to rise. Cool air from the polar regions then flows under the warm tropical air. These movements create air currents that circulate around the Earth. These are influenced by the Earth's rotation, the surface features of the continents and variations in the amount of moisture in the atmosphere. The Sun's rays warm the tropical waters, under which the cold polar wind flows and the wind forms the ocean currents.

We can make use of solar energy for heating water for domestic use, heating of buildings, drying agricultural products, generating electrical energy, etc. (Fig. 5.1). In the 1830s, the British astronomer John Herschel used a solar collector box to cook food during an expedition to Africa. Now, people are trying to use the Sun's energy for lots of other things. Electric utilities are trying photovoltaics, a process by which solar energy is converted directly to electricity. Electricity can be produced

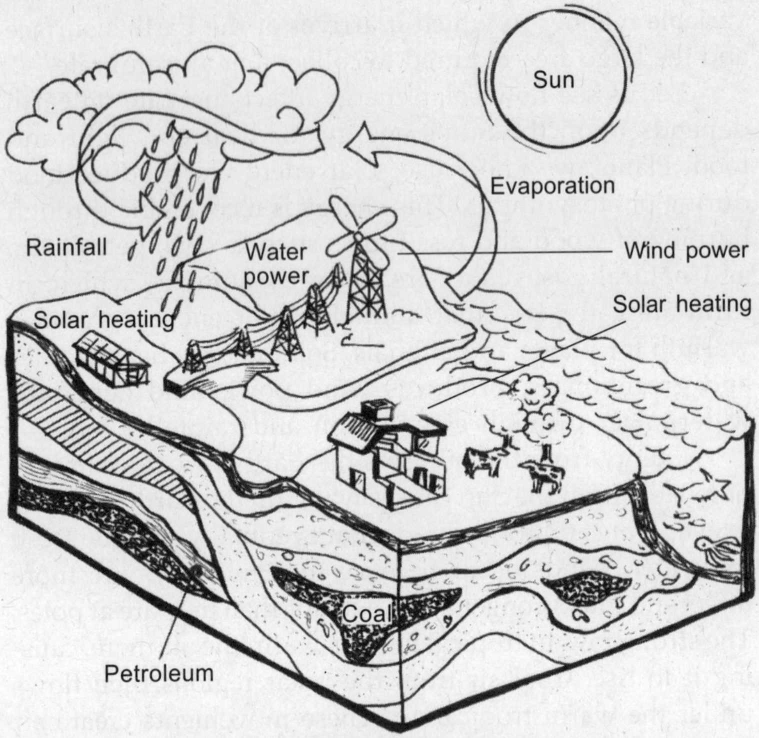

Fig. 5.1: How solar energy affects the Earth

directly from solar energy using photovoltaic devices or indirectly from steam generators using solar thermal collectors to heat a working fluid.

Solar Energy Benefits People

Solar energy has made news since the last fifty years. People today have become concerned about global warming and climate change and hence have also become aware of the uses of solar energy. A report by Shell Renewables, US, a division of one of the world's largest oil companies quotes that "By the year 2050, one half of the energy used worldwide will come from solar and other renewable sources". Back in 1952, a report prepared by the Paley Commission for US President Harry Truman predicted a bright future for solar energy. In the past few years, modern solar technologies have begun penetrating the market at faster rates.

Wind Energy

Man has harnessed the wind's energy for more than a thousand years. In places where the wind is steady and strong, wind farms or large group of windmills have been set up to generate electricity. Modern windmills used to produce electricity are often called wind turbines. Sunlight heats the ground and lower atmosphere produces the wind which powers wind turbines.

Wind turbines are mounted on a tower to capture most of the energy. At 200 m above the ground, they can take advantage of the faster and less turbulent wind. The three propeller-like blades catch the wind's energy. Each blade acts much like an airplane wing. When the wind blows, a pocket of low-pressure air forms on the downwind side of the blade. The low-pressure air pocket then pulls the blade toward it, causing the rotor to turn. This is called a 'lift'. The force of the lift is actually much stronger than the wind's force against the front side of the blade, which is called a 'drag'. The combination of lift and drag causes the rotor to spin like a propeller, and the turning shaft spins a generator to make electricity. Today, farms of wind turbines—called wind farms—are growing in number and helping generate electricity.

Ocean Energy

Ocean currents can have plenty of stored solar energy. The use of the ocean tides have been harnessed to make electricity along with a variety of other methods which make use of the motions and thermal gradients in the ocean. The temperature difference between the

Sun-warmed surface layers of the ocean and the colder depths through the use of a heat engine can derive useful energy, by a process called ocean thermal energy conversion (OTEC). This technology is complex, therefore limiting the use of the tremendous amount of stored energy in the ocean thermal gradients.

The Sun even works after dark. The gravitational pull from the Sun and the moon creates tides in the ocean, giving rise to tidal energy, which can be harnessed to make electricity. One way tidal energy is captured is by a dam that forces tidal water through turbines. As the water moves through the dam and turbines, the turbines turn and electricity is generated. Hydro energy is related to the Sun-powered cycle of water evaporation and rain. Hydroelectric power is renewable and considered a 'clean' energy since no burning is required; but it is limited in quantity.

Bio-energy

The solar energy is stored in plants and animals. This energy can be made use of in a variety of ways. For instance, trees can be burned as firewood. Plants or other organic matter, or 'biomass' can be burned in place of fossil fuels, such as coal or natural gas, to generate electricity. Fuel made from organic material can also be used as a substitute for gasoline or diesel fuel for transportation. Burning biomass produces less carbon dioxide than fossil fuel. Also, unlike fossil fuels, bio-energy is renewable. The raw materials, plants and trees, can not only be replaced, but as they grow, they help clean the air by removing carbon dioxide.

Using Sun's Energy Directly

How can we help stop global warming and slow down the harmful destruction of non-renewable resources?

Billions of tons of crude oil could be salvaged each year if we all did something simple like get a solar hot water system, which can save resources and prevent global warming.

Sun's energy can be directly harnessed for domestic purposes such as cooking water heating, cooling buildings and generating electricity. People have used the Sun's energy since ancient times. Greeks in 400 BC filled glass spheres with water for concentrating the Sun's rays to start fires. Greeks and Chinese also used curved mirrors to concentrate Sunlight for lighting fires.

The Anasazi Indians built dwellings with massive south-facing stone or adobe walls that absorbed heat during the day and released it at night. Horace Benedict de Saussure, a Swiss scientist, built the first hot box in 1767. It was a glass-covered wooden box with cork insulation to collect sunlight. In 1909, William J. Bailey, an American engineer, developed the first modern flat-plate collector. An American architect, George Fred Keck, built the first modern passive solar home in 1940. The south walls of the house were covered by windows made of two panels of glass with a thin layer of air sealed between them.

Solar Heating

Heating water on a gas stove or an electric stove can be cumbersome and also expensive. There is an alternative for this. Water can be heated in simple, inexpensive batch heaters in places with warm climate. A batch heater is an insulated tank with several layers of clear glass, covering the outside of the tank that faces south. The outside of the tank is blackened for maximum absorption of sunlight. The sunlight is converted to heat, which in turn warms the water inside the tank. The glass outside prevents most of the heat from escaping the tank. The hot water rises to the top and flows out of the tap.

At present, the general practice is to use flat-plate solar-energy collectors with a fixed position for heating swimming pools, heating water for domestic use, and space heating of buildings. A flat-plate collector consists of an insulated box covered by one or more layers of clear glass or plastic. Inside the box is a plate of black metal or black plastic. The plastic absorbs sunlight and converts it to heat, which gets trapped under the glass. Where space heating is the main consideration, the highest efficiency with a fixed flat-plate collector is obtained if it faces approximately south and slopes at an angle to the horizon equal to the latitude plus about 15 degrees.

Air, water or some other fluid circulates through tubes welded to the plate and absorbs heat from the plate. The heated fluid then flows to a heat exchanger, where it transfers its heat to the water. The heated water is stored in a tank and is pumped from the tank to the taps in the house.

Passive Solar Heating

We all have experienced passive solar heating of water. Turn on your taps in the afternoon during hot summer and you can feel the hot water on your hands. While you were not looking, the Sun was quietly working to give you hot water, even if you didn't want it. Well, if it's that easy, imagine what you can do if you are actually trying to make hot water. Passive solar hot water systems are probably the oldest commercially available solar systems.

Direct heating of a building by maximising the solar gain in the winter and minimising it in the summer is called passive solar heating. Many a times passive solar energy systems are used for heating air in buildings and houses in places with very cold climates. In most cases, these buildings have large south-facing windows to trap heat. During the day, sunlight passes through the windows and heats walls and the floors made of stone or brick. At the night, the walls and floors release the heat. Designs of northern homes and buildings use rock, water and other materials to store solar heat in the day to release it later at night.

Additional heat can be stored by placing water or special phase-change materials inside the walls. These materials melt at room temperature and as they melt, they store a large amount of heat and later release the heat as they become solid again. Special insulating shutters or shades also help keep the heat from escaping through the windows at night.

Active Solar Heating

Active space and solar heating is in contrast to passive solar heating. A water or air solar collector is used to heat a fluid that is used as the heating system for a building. In larger active thermal power generating systems, focused mirrors are used to concentrate and direct the sunlight into a boiler that produces steam to generate electricity.

Solar Air-conditioning

Producing heat using sunlight sounds logical, but solar air-conditioning sounds like fiction, doesn't it? Solar air-conditioning makes use of solar collectors and special materials called 'desiccants' that can absorb large amounts of water. The cooling begins when fans force air from outside through a desiccant, which removes moisture from the air. The air then flows through a revolving wheel that acts as a heat exchanger and removes heat. Next, the air passes over a surface soaked with water. As the water comes in contact with dry air, it evaporates and absorbs more heat from the air. The air cools the building and as it leaves the building, the solar collectors reheat it. Blowing the reheated air through it dries out the desiccant and the process restarts.

Generating Electricity Using Solar Energy

Direct solar energy can be used to create electricity, using devices such as photovoltaic cells or 'solar cells' and solar furnaces. Today we have solar calculators that use solar cells to power them.

Photovoltaic Cells

An immense amount of energy from the Sun strikes the surface of the Earth every day. This energy may be captured and used in the form of heat in 'solar thermal' applications, or it may be converted directly into electricity, to power electrical devices using photovoltaic cells (Fig. 5.2).

Photovoltaics have a great impact on our lives. These cells work silently, not polluting yet generating energy

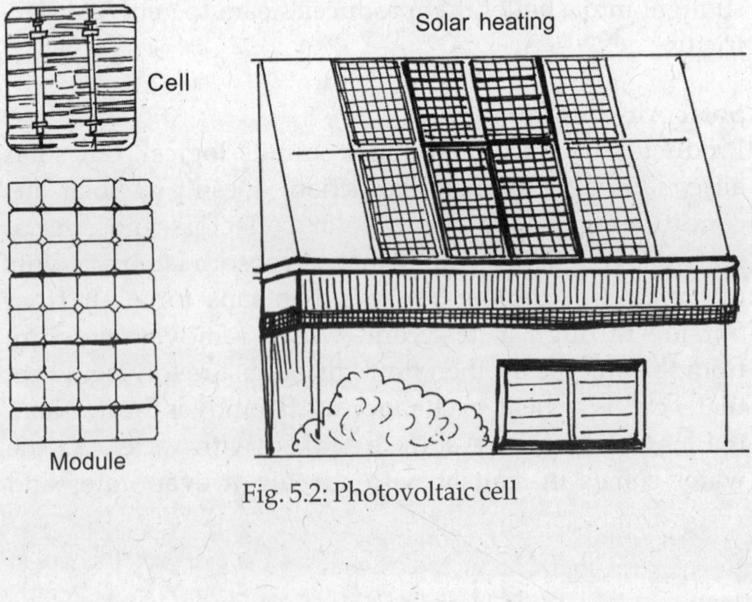

Fig. 5.2: Photovoltaic cell

whenever the Sun shines. A photovoltaic cell is a non-mechanical device usually made from silicon alloys. It was invented in 1954 by engineers at Bell Telephone Laboratories in the USA, who examined the sensitivity of a properly prepared silicon wafer to Sunlight. A photovoltaic cell consists of thin slices of semi-conductor materials. Sunlight is composed of photons, or particles of solar energy. These photons contain various amounts of energy corresponding to the different wavelengths of the solar spectrum. When photons strike a photovoltaic cell, they may be reflected, pass right through, or be absorbed. Only the absorbed photons provide energy to generate electricity. Photovoltaic cells, like batteries, generate direct current (DC), which is generally used for small electronic equipment.

Individual cells can vary in size from about 1 cm to about 10 cm across. However, one cell produces only 1–2W, which is not enough power for most applications. To increase power output, cells are electrically connected into a packaged weather-tight module. Modules can be further connected to form an array. An array of such cells can be used to power electronic equipment. The performance of a photovoltaic array is dependent upon Sunlight. Climatic conditions such as clouds or fog have a significant effect on the amount of solar energy received by a photovoltaic array and, in turn, on its performance. Most current technology-photovoltaic modules are about 10 per cent efficient in converting Sunlight to electricity with further research being conducted to raise this efficiency to 15 per cent.

Photovoltaic cells power most artificial satellites, wrist-watches and electronic calculators. Without telecommunications satellites, many of our now-routine activities—from watching internationally broadcast entertainment, news, etc. to using cell phones—would still be in the realm of science fiction. And space exploration and research too might still be science fiction. More complicated systems provide electricity to pump water, power communications equipment, and even provide electricity to our homes. In rural areas, photovoltaic technology is significantly improving living conditions. It can offer unique opportunities to improve rural health care, education, agriculture and lighting. Photovoltaic systems are even being incorporated in the building of roofs and walls of buildings in the urban areas.

Solar Furnaces

These are high-temperature collectors that generate large amounts of electricity. These can include many flat or slightly curved mirrors that focus the Sun's rays on a target, such as a piece of metal. For instance, a solar thermal power plant uses hundreds of mirrors to concentrate and reflect the Sun's rays. The curved mirrors focus Sunlight into a cylinder. Pipes carry water into the cylinders. The reflected Sunlight warms the water to produce steam, which is converted into mechanical energy, which drives the turbines to generate electricity.

Solar thermal power generation is essentially the same as conventional technologies except that in conventional technologies, the energy source is from the stored energy in fossil fuels, which is released by combustion. Solar thermal technologies use 'concentrator' systems due to the high temperatures needed for the working fluid. The three types of solar-thermal power systems in use or under development are—parabolic trough, solar dish, and solar power tower.

Parabolic Trough

The parabolic trough is the most advanced of the concentrator systems. A parabolic trough collector has a linear parabolic-shaped reflector that focuses the Sun's radiation on a linear receiver located at the focus of the parabola. The collector tracks the Sun along one axis from east to west during the day to ensure that the Sun is continuously focused on the receiver. Because of its parabolic shape, a trough can focus the Sun at 30 to 100 times its normal intensity (concentration ratio) on a receiver pipe located along the focal line of the trough, achieving operating temperatures over 400ºC.

A collector field consists of a large field of single-axis tracking parabolic trough collectors. The solar field is modular in nature and is composed of many parallel rows of solar collectors aligned on a north-south horizontal axis. A working fluid is heated as it circulates through the receivers and returns to a series of heat exchangers at a central location where the fluid is used to generate high-pressure superheated steam. The steam is then fed to a conventional steam turbine/generator to produce electricity. After the working fluid passes through the heat exchangers, the cooled fluid is re-circulated through the solar field. The plant is usually designed to operate at full rated power using solar energy alone, given sufficient solar energy. However, all plants are hybrid solar/fossil plants that have a fossil-fired capability that can be used to supplement the solar output during periods of low solar energy.

The technology is used in the largest grid connected solar-thermal power plants in the world. One such complex in the US uses parabolic troughs. The Kramer Junction companies operate and maintain five 30-megawatt Solar Electric Generating Systems (SEGS). These SEGS comprise 150 to 354 MW of installed parabolic trough solar thermal electric generating capacity located in California's Mojave Desert. The combined California facilities produce more than 90 per cent of the world's commercially available solar thermal electric power.

Solar Dish

A solar dish/engine system utilises concentrating solar collectors that track the Sun on two axes, concentrating the energy at the focal point of the dish because it is always pointed at the Sun. The concentration ratio of a solar dish is much higher than the solar trough, typically over 2,000, with a working fluid temperature over 750ºC. The power-generating equipment used with a solar dish can be mounted at the focal point of the dish, making it well suited for remote operations or, as with the solar trough, the energy may be collected from a number of installations and converted to electricity at a central point.

The engine in a solar dish/engine system converts heat to mechanical power by compressing the working fluid when it is cold, heating the compressed working fluid, and then expanding the fluid through a turbine or with a piston to produce work. The engine is coupled to an electric generator to convert the mechanical power to electric power.

Solar Power Tower

A solar power tower or central receiver generates electricity from Sunlight by focusing concentrated solar energy on a tower-mounted heat exchanger (receiver). This system uses hundreds to thousands of flat Sun-tracking mirrors called 'heliostats' to reflect and concentrate the Sun's energy onto a central receiver tower. The energy coming from the Sun can be concentrated as much as 1,500 times. Energy losses from thermal-energy transport are minimised as solar energy is being directly transferred by reflection from the heliostats to a single receiver, rather than being moved through a transfer medium to one central location, as with parabolic troughs.

Power towers must be large to be economical. This is a promising technology for large-scale grid-connected power plants. Though power towers are in the early stages of development compared to the parabolic trough technology, yet a number of test facilities have been constructed around the world.

Solar One, near Barstow, California, US, which operated from 1982 to 1988, at about 10 MW, was the world's largest power tower plant. In Solar One, water was converted to steam in the receiver and used directly to power a steam turbine. The heliostat field consisted of approximately 1,800 heliostats. The storage system stored heat from solar-produced steam in a tank filled with rocks and sand using oil as the heat-transfer fluid. A consortium comprising the US Department of Energy and a number of electric utilities, led by Southern California Edison, redesigned Solar One to a more advanced molten-salt technology—Solar Two, which started operation in 1996.

Solar Cooking

You can now cook your food using parabolic dish-shaped reflectors. These focus Sunlight on the food or on a pot that contains food. You can even use a solar oven to cook your food. This consists of an insulated box with a window and several reflective surfaces. The oven heats up when the window is pointed toward the Sun (Fig. 5.3).

Solar Food Drying

We in India eat a lot of Sun dried food. Sunrays and the heat act as preservatives for potato wafers, sago wafers, a variety of *papads*, pickles, *chunda*, etc. Dried *methi* or fenugreek leaves are commonly used. A variety of mouth fresheners such as *saunf, til, ajwain*, etc. are all Sun dried. The need for solar dryers in areas where fruit is plentiful

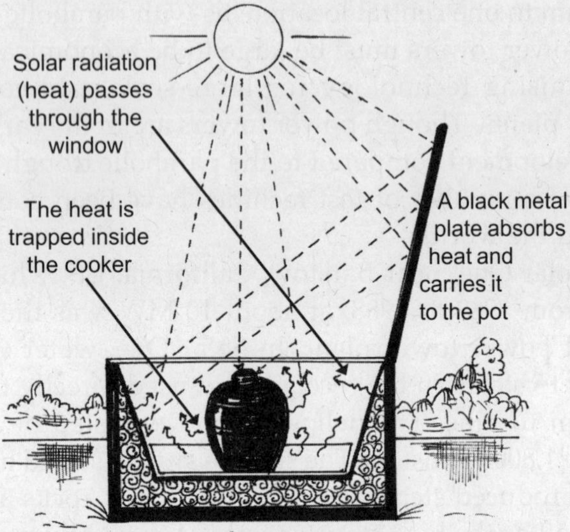

Fig. 5.3: Solar cooker

in summer months is great, but because there is no simple and economic method to preserve it, much of it is left to rot. But the art of drying food, especially fruits and vegetables using solar energy is a little more complicated than you might think. Solar food drying is affected by many variables, especially the amount of Sunlight and relative humidity. There are some basic guidelines to drying vegetables and fruits.

Wash fresh fruits and ripe vegetables thoroughly. Effective drying is accomplished with a combination of heat and air movement that removes 80 to 90 per cent of moisture from the food. Typical drying times range from 1 to 3 days, again depending on Sun, air movement, humidity, and type of food. Once the drying process has started it should not be interrupted. Direct Sunlight is not recommended. Too much heat especially early in the process will prevent complete drying. Fruits or vegetables should be cut into thin slices, less than 1 cm thick and spread out on trays to allow free air movement. Rotate trays 180 degrees daily for uniform drying. Move dryer food to bottom racks. Safe tray materials include stainless steel rack, wood slats, cheesecloth, Teflon, Teflon coated fiberglass, nylon, food grade plastics, etc.

Allow food to cool completely before storing. Store food in airtight jars or containers, and do not expose dried food to air, light or moisture. Most dried fruits taste great and this includes apples, apricots, bananas, grapes, etc. Vegetables are best reconstituted by covering with cold water until they are near original size. They can be added in their dry form to soups/stews. Vegetables can also be ground into powders and used for instant soups or flavoring.

Solar Energy and Agriculture

Crops too depend on the correct amount of Sunlight and rain. The moment this balance is disturbed, the crops will be lost.

Solar energy can be used in agriculture in a number of ways, which can save money, increase self-reliance, and reduce pollution. Whether drying crops or powering a water pump, the Sun can make the farm-working efficient. Use of solar energy can begin by designing barns and sheds to use natural light, instead of electric light. The Sun's heat can be used to warm the livestock buildings. Solar water heaters can provide low- to medium-temperature hot water for cleaning equipment and to warm and stimulate cow's udders.

Using the Sun to dry crops and grain is one of the oldest applications of solar energy. Solar drying equipment can dry crops faster and more evenly than by leaving them in the field after harvest, with the added advantage of avoiding damage by birds, pests, and weather.

Solar energy can be of great use in maintaining

constant temperatures at commercial greenhouses. In order to capture the maximum Sunlight, a solar greenhouse should face south, and the north side should be well insulated, with few or no windows.

For generating electricity around the farm, photovoltaic (PV) panels are often a cheaper option than new electric lines for providing power to remote locations. Also, as they do not require any fuel and have no moving parts, they are convenient to operate and maintain.

Newer Solar Technologies

Solar energy today is used in building sophisticated solar-powered facilities ranging from home to an industrial or commercial complex. A 354 MW solar power plant was built in the 1980s in the Mojave Desert, California, USA. Here the reflecting troughs concentrate the heat from the Sun's rays and raise the temperature to about 400ºC producing steam that runs a power generator. When the Sun is not shining, the plant switches to natural gas. Improvements and modifications in such plants have been made today using new engineering solutions and scientific principles such as non-imaging optics that help build efficient heat concentrators at a lower cost.

The newer solar technologies play an important role in energy-saving projects. Buildings can be the largest collectors of solar energy. We need energy to light up our homes, get hot water and run a number of devices. The newer building technology employs energy related elements such as insulating materials to high-performance windows, special glass, solar-powered heating, modified roofs and walls or low consumption light bulbs. Architects are tackling the building process with great integration keeping in mind the concept of energy saving and use of solar energy, especially using the active and passive solar systems.

Tremendous progress has been made in the development of modern solar technologies such as using solar wind and biomass for the production of fuel, heat and electricity. These are already providing services to individual homes, villages and cities.

The Indian Scene

In India, the photovoltaic generating capacity at the end of the year 1999 was 44,000 kWp according to the data available from Trends in Photovoltaic Applications, September 2000, IEA Photovoltaic Power Systems Programme; Ministry of Non-Conventional Energy Resources, Government of India.

The Indian Renewable Energy programme is constituted under the Department of Science and Technology before being transferred to the newly created Department of Non-Conventional Energy Sources in 1982. The Department was upgraded to the Ministry of Non-Conventional Energy Sources (MNES) in 1992 and MNES has since worked with the Indian Renewable Energy Development Agency (IREDA—created in 1987), to accelerate the momentum of renewable energy development. The promotion has been achieved through R&D, demonstration projects, government subsidy programmes and programmes based on cost recovery supported by IREDA and also private sector projects.

India receives a good level of solar radiation, the daily incidence ranging from 4 to 7 kWh/m^2 depending on location. The Solar Energy Programme that is being implemented by the MNES encompasses both solar thermal and solar photovoltaic technologies. The Programme, regarded as one of the largest in the world, plans to utilise India's estimated solar power potential of 20 MW/km^2 and 35 MW/km^2 solar thermal. The country has also developed a substantial manufacturing capability, becoming a lead producer in the developing world.

The principal objective of the Solar Thermal Programme is the market development and commercialisation of solar water heaters, solar cookers, etc. At the present time, the installed systems account for some 500,000 m^2 collector area and some 485,000 solar cookers. Solar water heating has been used by individual homes, hotels and industrial processes. The near-future potential for such systems is around 30 million m^2 of collector area. Solar air heating has been utilised in various parts of the country for drying agricultural produce and in timber kilns. Solar stills have been employed in large numbers to supply distilled water in rural hospitals, battery-charging stations and for the supply of drinking water in remote arid zones.

The MNES has been promoting the sales of box solar cookers since the early 1980s. This type of cooker is designed to prepare food for up to 4-5 people and can be supplied with or without electrical back up. However, the Dish Solar Cooker designed for 10-15 people and the Community Solar Cooker for 35-40 people have also been developed. In March 1999, the world's largest Solar Steam Cooking System was installed at Mount Abu, Rajasthan. It is a hybrid system with back-up oil-fired boilers and is designed to prepare food for 10,000 people.

A Solar Buildings Programme is aimed at creating an awareness of the potential for solar-efficient buildings. The passive solar design concept is a climate-responsive architectural practice that is now being researched, developed and implemented throughout the country. During 1999, a proposal for a 140 MW integrated solar combined-cycle power project with a solar thermal power capacity of 35 MW was sanctioned. The plant, based on the parabolic trough collector technology, is located in the Jodhpur district of Rajasthan and will have supplementary firing by naphtha gas on Sunless days.

MNES has also developed a solar PV programme aimed particularly at rural and remote areas. This includes street lighting and solar lanterns, home lighting, water pumps, telecommunications, power plants and other applications. The MNES has instituted a plan for establishing solar PV power generation for use in specialised applications such as voltage support at rural sub-stations and peak saving in urban centres.

Communication

The light of the Sun has been used by man throughout the ages to communicate with each other. According to the Greek historian Herodotus, Greeks and Romans used polished mirrors or polished combat shields that reflect light to send signals. This was known as the 'Heliograph signaling'. Ships and armies on a battlefield used to communicate using light flashes. In the 1880s US Army Troops used mirrors to communicate between mountain ranges in the western part of the United States. Even today we 'harness the Sun' to communicate. Radio signals are used a great deal to communicate and solar activity is constantly bombarding the ionosphere of Earth and positively affecting communications. This has far reaching repercussions on anyone or anything that is dependent on the transmission of signals.

Radio-propagation

The solar activity of the Sun influences radio-wave propagation by affecting the Earth's ionosphere. The ionosphere consists of four layers of charged particles, which are ionised by ultraviolet radiation from the Sun. The layers have been given letter names starting with the D layer that is the lowest layer that affects radio-wave propagation. The layers extend out to the E, F_1, and F_2 layers with the F_2

layer most responsible for long-distance amateur communication. There are no A, B, or C layers. The ionosphere refracts or bends radio waves back to the Earth. Day-to-day and yearly solar conditions influence the radio signals, which makes radio-wave propagation difficult to predict. Maximum ionisation occurs during peak Sunspot activity. High Sunspot numbers at the cycle peak normally produce the best wave propagation conditions.

Solar flux or radio energy coming from the Sun produces greater ionisation in the ionosphere. Large antennas pointing at the Sun measure solar flux, which may be taken under any weather conditions. The solar flux index number along with the Sunspot number that are both directly related with the activity of the Sun is used to predict radio-wave propagation. Flux values between 60s and 70s mean fair to poor propagation.

Solar energy or power from the Sun is clean and unlimited. Capturing the Sun's energy and taking advantage of it for light, heat, hot water, and electricity can be a convenient way to save money.

6

SUN AND YOU

You know that you can't fly directly to the Sun. For us to fly anywhere near the Sun we would have to build a spacecraft that could withstand heat over 1,000,000 degrees Kelvin, keep the people inside protected from not only the extreme heat, but also the intense light and solar radiation. Even at this present distance, too much exposure to the Sun can be dangerous.

Greenhouse Effect
The light of the Sun warms the Earth, but the heat it creates cannot easily pass through the atmosphere into

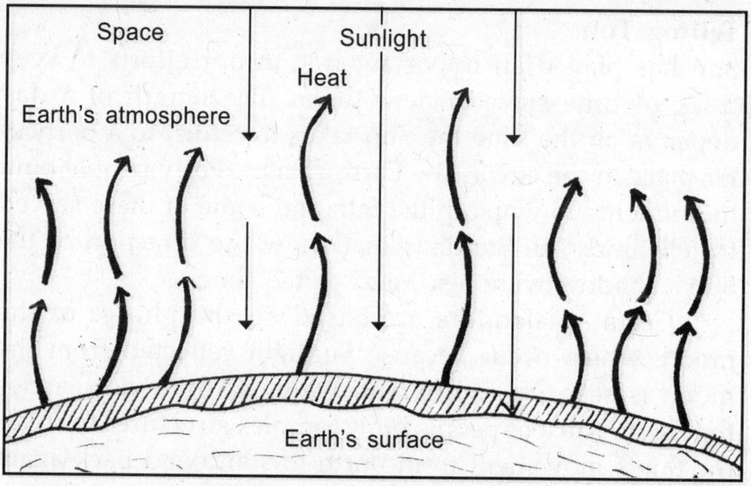

Fig. 6.1: The greenhoue effect

space due to which the Earth gets warmed by the Sun. This behaviour of the atmosphere is called the greenhouse effect because it resembles the action of a 'greenhouse' (Fig. 6.1). A greenhouse lets the Sunlight in to heat the plants, but the heat passes back through the roof and walls very slowly. Planting of trees can help trap the heat of the atmosphere. This is one way in which we can help maintain the atmospheric temperature. The Greenhouse Effect is essential to life on Earth because without it temperatures would be too low to support life, as we know it.

Ozone Layer

Ozone is one of the layers of Earth's atmosphere that acts as a kind of sunscreen protecting living organisms from harmful UV (ultraviolet) radiation emitted by the Sun. Unfortunately, today there is a thinning of ozone layer due to air pollutants such as chlorofluorocarbons (CFCs) given off by the cooling systems of refrigerators, air-conditioners etc. These destroy the ozone layer.

Telling Time

Sun has played an important role in our efforts to keep track of time since ancient times. The length of a day depends on the time the Sun takes to return to a particular place in the sky as the Earth rotates. Early people built monuments to worship the Sun and some of these served to tell time. The Sundials that show the direction of the Sun's shadow were also used to tell time.

Certain calendars are based on the phases of the moon, which occur because Sunlight reflected from the moon is seen from different angles as the moon circles the Earth. Ancient people built complex structures to study the Sun's movement from north to south and back again as the seasons changed. Monuments built in Stonehenge in England tracked the motions of the moon and the Sun at that time. Even today the Sun helps in navigating and surveying.

Sunshine Cheers You Up

A sunny day can alter your mood chemically and even prevent depression. Seasonal Affective Disorder (SAD), also known as 'cabin fever', strikes when there is less sunlight, especially during the winter months. SAD is a form of clinical depression. Scientists suspect that the hormone Serotonin is responsible for SAD because it influences brain activities including sleep and eating behaviours, and mood regulation.

People with SAD experience depression, fatigue, lack of energy, increased need for sleep, craving for carbohydrates, intense sadness, sudden weight gain, and craving for sweets. The onset of spring gives thousands of people relief from SAD. These individuals can be treated with exposure to light. Exercise can be helpful. Medications such as Tryptophan can boost the effectiveness of light treatment. A balanced diet comprising more vegetables, fruits, and complex carbohydrates will also help. Sugars, processed or junk foods should be avoided. Remember SAD is a psychological disorder and requires professional medical diagnosis and treatment.

Sunlight also stimulates the pineal gland—a tiny pea-sized organ found in the base of the brain (Fig. 6.2).

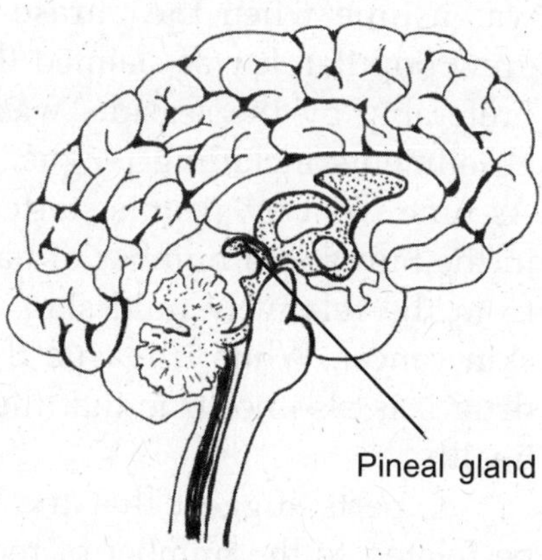

Fig. 6.2: Pineal gland

Sometimes known as 'the third eye', the pineal gland produces certain types of brain chemicals called 'tryptamines'. One type of tryptamine, melatonin, keeps our body clock aware of day and night, and the changing seasons.

The Sun may offer a cheaper alternative to traditional medical lasers, as in the report by Israeli researcher Jeffrey Gordon, Ben-Gurion University, Negev, Israel. Laser therapy for cancer involves the use of high-intensity light to destroy cancer cells. The Sun offers hope for a cheaper cancer cure. The system uses a collector outside the laboratory window. A mirror gathers Sunlight and the flat mirror above the dish sends the solar energy through a fiber optic cable in the laboratory's floor. This could make the solar operated lasers more affordable for hospitals than the traditional lasers.

Our Health and Sun

One of the major areas where the Sun's contribution is often discounted is in its promotion of health. From the very beginning of this century people have considered the Sun as being good for health. In the early 1920s, the Sun was touted as a cure for all major diseases. There was a time when the phrase 'recuperating in the Sun' grew popular. It was claimed that a brief spell in the Sun, preferably by the seaside, was a kind of magic cure for all ailments including plague, old age, and tuberculosis. As time went by, people started spending too much time in the Sun and a number of harmful effects were noticed from the relatively mild skin rashes to the more serious skin cancer. While it is true that the Sun isn't a wonder drug it is also not true that the Sun does nothing for our health.

Experts suggest that the influence of Sunlight may be related to the number of red cells and haemoglobin in the blood. An insufficiency of light may cause an increase in the serum or watery portion of the blood and a corresponding decrease in the quantity of blood fibrin and red corpuscles, resulting in anaemia. Sunshine could even help in increasing the coagulating power of the blood, and thus may help those who suffer from uterine haemorrhage. Sunlight may also favourably influence the irregularities of ovulation, pubertal difficulties and impotency. A Sunbath may aid in some neurological affections. Studies have shown that after prolonged fasting or a wasting illness, exposure to sufficient Sunshine may enable the body to gain energy.

Scientists today also believe that sunlight may reduce the risk of several types of cancer such as breast, colon, ovary, bladder, womb, stomach and prostate gland.

Vitamin D

The Sun plays a very important role in the formation of vitamin D in our body. When Sunlight penetrates our skin, vitamin D is produced. Vitamin D is called the 'Sunshine vitamin' because it is formed in the skin when the body is exposed to Sunlight. This vitamin is essential for our health. A major source of this vitamin is produced through the interaction of the Sun with certain chemicals in our body.

Vitamin D is essential for calcium and phosphorous metabolism in our body. It helps prevent rickets. The deficiency as well as the excess of this vitamin can seriously affect the bones. Amongst the several forms of vitamin D, calciferol or vitamin D_2 is the one that is produced in plants. Cholecalciferol or vitamin D_3 occurs in tissues of animals, including human beings. Sunlight is the primary trigger for vitamin D, but it is also found in foods such as cheese, butter, margarine, etc. (Fig. 6.3)

The latest research suggests that the chance of developing schizophrenia may be directly linked to how

Fig. 6.3: Sources of vitamin D

sunny it was in the months before you were born. A lack of sunlight can lead to vitamin D deficiency, which scientists believe could alter the development of a child's brain in the womb. People who develop schizophrenia in Europe and North America are more likely to have been born in the spring, according to an article in the *New Scientist*.

Ten minutes of daily exposure to sunlight can supply us with all the vitamin D that we need. People with darker skin need more exposure than light-skinned people. John McGrath, a psychiatrist at the Queensland Centre of Schizophrenia Research in Brisbane, Australia, suggested that a lack of UV light during pregnancy tips the balance towards schizophrenia in genetically susceptible people.

Low vitamin D is associated with several autoimmune diseases including multiple sclerosis, rheumatoid arthritis, thyroiditis and Crohn's disease. Recent laboratory experiments suggest that vitamin D can also prevent the growth and spread of cancerous tumours.

Bones

Sunlight is vital for proper bone development. This is due to the fact that Sunshine, particularly the ultraviolet rays during the growing years help lay down and fix the calcium and phosphorous salts in an ideal fashion so as to make for the transformation of cartilage into bone. On the other hand, when there is insufficient exposure to sunlight, then defective, misshapen, and brittle bones may result. Sunlight helps in the assimilation of calcium and because of that it is of great value in the prevention of rickets and tuberculosis (Fig. 6.4). A lack of calcium is associated with both these conditions.

Skin

The Sun is said to maintain the general health of our skin and acts like a natural disinfectant for the body. Tanning is your skin's natural response to UV light. It is a reaction, just like an allergic reaction, to prevent further injury to your skin from the Sun though it does not prevent skin cancer. Acne, a glandular disturbance of the skin, is also noticeably aided by Sunlight, as is the skin condition called 'psoriasis'.

Too much Sunlight can be harmful; it can burn the skin especially in hot and humid weather. Sunburns can be mild and severe. The skin turns red in mild Sunburn and the redness disappears in a few days. Severe Sunburns on the other hand can produce blisters on the skin and may even be accompanied by chills, dizziness and fever. Repeated Sunburn over a long period of time may cause skin cancer and excessive wrinkling.

Fig. 6.4: A boy with rickets

It is important to know which kind of skin you have because in this way you're able to individuate the possible risks. Burning of skin depends in particular upon the colour of your skin and hair colour. Thankfully, we Indians have dark hair and dark skin, and are the ones at minor risks. People who have this type of skin might expose themselves to the Sun without any particular attention and they will obtain a quick, homogeneous and perfect tanning.

Sunburns can be avoided by the use of Sunscreen lotions and creams, which block the Sun's burning rays, or by gradual exposure to the Sun, which results in Suntan. But repeated Sun tanning may also contribute to skin cancer and wrinkling. Sunscreens come in many Sun Protection Factors (SPFs), from 2-45. An SPF is the amount of protection that you get from Sunscreens. Use a Sunscreen with a higher SPF number, if you want to stay longer in the Sun. Higher the protective-number better will be the protection.

How Do Sunscreens Work?

The Sunscreen ingredients absorb or reflect ultraviolet radiation (UV) by forming a barrier on the skin surface. Energy, in the form of Sunlight, is converted to heat when it comes in contact with the Sunscreen, which then is dissipated. You can block UV rays with opaque creams like the white zinc oxide cream that you see lifeguards putting on their noses. You can also absorb UV radiation in much the same way that melanin does. The first and most common of the absorption chemicals is PABA (para-aminobenzoic acid), which absorbs ultraviolet-B radiation (UVB). Other chemicals include Cinnamates (absorb UVB), Benzophenones (absorb UVA) and Anthranilates (absorb UVA and UVB).

The SPF acts like a multiplying factor. If you would

normally be all right in the Sun for 10 minutes and you apply a Sunscreen with SPF 10, you will be fine in the Sun for 100 minutes. In order for the Sunscreen to work, however, you have to apply plenty and it has to stay on. Use a product, which is water-resistant but also in these cases it is better to use the cream more than once per day. You should apply it about half an hour before going out in the Sun (or the water) so it can bind to your skin because if you don't, then it is very easy for the Sunscreen to be washed off.

Skin Cancer

Long-term effect of too much Sun exposure is not completely predictable. There are, of course, other determining factors, including your heredity and the environment you live in. You may be at a higher risk of skin cancer if there is a history of skin cancer in your family, you have fair skin, or you have a northern European heritage. The elevation, latitude, and cloud cover of the area you live in effect susceptibility to skin cancer. UV light gets stronger as the elevation increases. The thinner atmosphere at higher altitudes does not filter UV as much as it does at lower altitudes. The Sun's rays are also stronger near the equator. However, both the total amount of Sun received over the years, and overexposure resulting in Sunburn can cause skin cancer.

Each time our skin becomes tanned or burned, damage is done to individual skin cells and DNA. Some cells die and some repair themselves by getting rid of the damaged DNA. Cells that cannot repair themselves eventually become defective cancerous cells.

Ultraviolet light contained within Sunlight lowers the body's immune system, and makes it difficult to destroy defective cells. These slowly grow to produce a cancerous tumour.

Skin cancer develops slowly. The damage to the skin may show up as one of three types of skin cancer: basal cell, squamous cell, or malignant melanoma. The skin has two main layers and several kinds of cells. The top layer of skin is called the epidermis. It contains three kinds of cells: flat, scaly cells on the surface called *squamous* cells; round cells called *basal* cells; and cells called *melanocytes*, which gives skin its colour.

Melanoma

Sometimes, a dark-pigmented malignant mole or tumour called 'melanoma' forms on the skin. Melanomas can appear suddenly with no warning or can develop from or around moles. It is very important to notice any changes in the number and appearance of moles on the body. Melanomas can occur anywhere on the body but are most frequently found on the upper backs and legs.

Melanoma is the most severe form of skin cancer, and it is the one most likely to spread to other parts of the body. The risk of developing one of these cancers increases with each episode of Sunburn throughout life. Knowing how to spot these cancers early can lead to timely, effective treatment. When checking your skin for cancer, look for fleshy bumps that bleed, scab over, and heal in cycles; scaly patches of skin; and moles that appear suddenly or begin to grow and change colour.

Babies less than six months old should be kept covered up and out of the Sun completely. At that age their skin does not produce enough melanin to protect them from UV light. Many dermatologists believe there is a link between childhood Sunburn and malignant melanoma later in life.

Photo Ageing

As you grow old, your skin loses its youthful appearance.

But the skin ageing depends upon the amount of exposure of Sunlight you have had during the course of life. The more the exposure, the more signs of ageing will show. The lesser exposure the more youthful a person's skin will remain. This is called photo ageing—ageing related to Sun. The ultraviolet rays (electromagnetic radiation) causes photo (light) damage. This occurs especially when the skin remains unprotected.

Photo damage can last for short term or can even last for years together. Short-term photo damage includes Sunburn and Suntan and changes in the skin's immune system. Long-term photo damage can cause photo ageing making the skin look old. Photo ageing causes wrinkles, dryness and roughness of the skin, loss of skin elasticity, change in the skin colour, appearance of dark freckles, easy bruising sets in and the skin becomes unable to protect itself against skin cancer. The worst effect of long-term photo damage is pre-cancer and cancer.

Photo ageing is more common amongst fair-skinned people whose skin generally tends to burn and not tan. Prolonged exposure to Sun can also cause photo ageing in dark as well as wheat-skinned people. The 'Sun' ageing shows mostly on the exposed parts to the Sun, such as face, ears, and neck, hands, forearms and perhaps a bald head. The best way to avoid photo damage and photo ageing is avoiding Sunlight when the Sun is directly overhead and it's intensity is at its highest. Topical ointments and creams help. However, it is difficult to reverse the damage that is caused by long-term photo damage.

Certain cosmetic procedures such as derma-abrasion and chemical peels are used to treat the sunburnt skin. Laser resurfacing is the latest procedure used to take care of the problem. In this the top layers of the skin are removed for the new healthy skin to take its place.

Sun Allergy

Some people are more sensitive to Sunlight than others. The most common reaction of all Sun allergies is called polymorphic light eruption (PMLE). Often confused with prickly heat, PMLE appears as small red, itchy eruptions on the skin. A rash can occur even from exposure to Sun through windows. This allergy is more common in adult women than men, usually appearing from late teens to 40s. It can affect all skin types, but is more common in fair-skinned individuals.

A skin type is classified according to the way it reacts to Sunrays.

Skin Phototype I: constitutes a person with very fair complexion with red or blond hair. This skin type is very sensitive to ultraviolet light and burns easily. People who have skin phototype I, have the greatest chance of photo ageing and skin cancer.

Skin Phototype II: constitutes a person with fair complexion with light or dark eyes. This skin type is very sensitive to ultraviolet light and burns easily. People with skin phototype II are also in the highest risk category for photo ageing and skin cancer.

Skin Phototype III: includes a lightly pigmented individual whose skin is rather sensitive to ultraviolet light and does not burn as easily as types I and II but instead gradually tans. People with Skin Phototype III are also in the high-risk category for photo ageing and skin cancer but are not as susceptible as those with types I and II.

Skin Phototype IV: includes a moderately pigmented individual whose skin is moderately sensitive to ultraviolet light but rarely burns. People with skin phototype IV have slightly less chance of photo ageing and skin cancer.

Skin Phototype V: includes a moderately pigmented individual whose skin is minimally sensitive to ultraviolet light that does not burn but always tans. People with skin phototype V have much less chance of photo ageing and skin cancer although it is still a possibility.

Skin Phototype VI: includes a heavily pigmented individual whose skin is not sensitive to ultraviolet light at all. Such individuals always tan and never burn however they are still susceptible to photo ageing and skin cancer.

Too Much Ultraviolet (UV) Light

It is true that in the early part of this century the Sun was considered as a 'magic cure' for all major diseases. People flocked to the seaside to spend some time in the Sun to ensure their good health. At one point people considered the Sun a cure even for fatal diseases like tuberculosis. The Sun does play a very important place in the general health of the body but is not really a wonder drug. Like anything in excess, even the Sun, may be harmful to health. In fact, it is the amount of Sunlight and the degree of exposure that is the critical factor.

Stratospheric ozone protects the biosphere from potentially damaging doses of UVB (Fig. 6.5). However, the

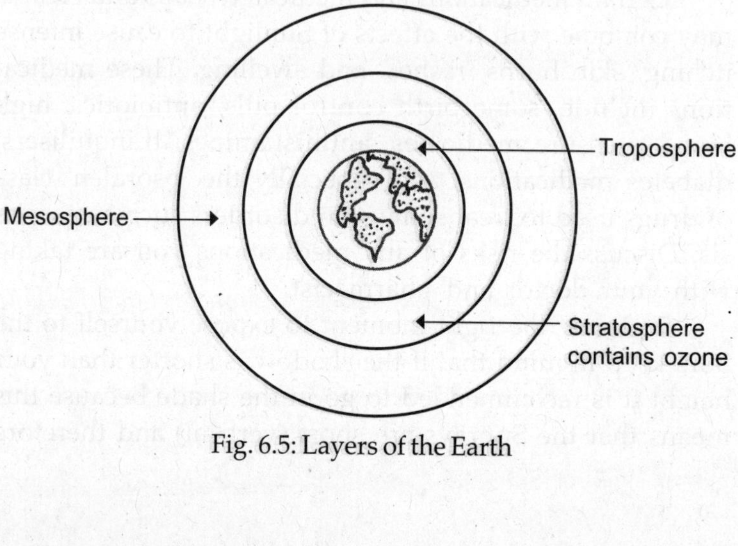

Fig. 6.5: Layers of the Earth

recent depletion of stratospheric ozone could lead to significant increases in UVB reaching the Earth's surface. UVB radiation is responsible for a wide range of potentially damaging human and animal health effects, primarily related to the skin, eyes, and immune system. Exposure to UVB depends upon an individual's location (latitude and altitude), the duration and timing of outdoor activities (time of day, season of the year), and precautionary behaviour (use of Sunscreen, Sunglasses, or protective clothing). An individual's skin colour and age can also influence the occurrence and severity of some of the health effects from exposure to UVB.

Try and avoid intense exposure to the hot Sun. Full-spectrum Sunlight, the way most of us 'catch a few rays', includes both ultraviolet A and B light (UVA and UVB), which cause tanning and burning of the skin. The UVA radiation goes deeper into and beneath the skin, its effects may not be as apparent until damage is done. Ultraviolet light can cause problems in all skin types depending on the degree of exposure. Generally, if Sun causes any redness or discomfort, it means too much exposure (Fig. 6.6).

Certain medications and medicated soaps and creams may combine with the effects of Sunlight to cause intense itching, skin burns, rashes, and swelling. These medications include, some birth control pills, antibiotics, high blood pressure medicines, antihistamines, tranquilisers, diabetes medications, and especially the 'psoralen' class of drugs used to treat some skin disorders such as psoriasis. Discuss the risks of any medications you are taking with your doctor and pharmacist.

To know the right moment to expose yourself to the Sun, keep in mind that if the shadow is shorter than your height it is recommended to go in the shade because this means that the Sunrays are short (vertical) and therefore

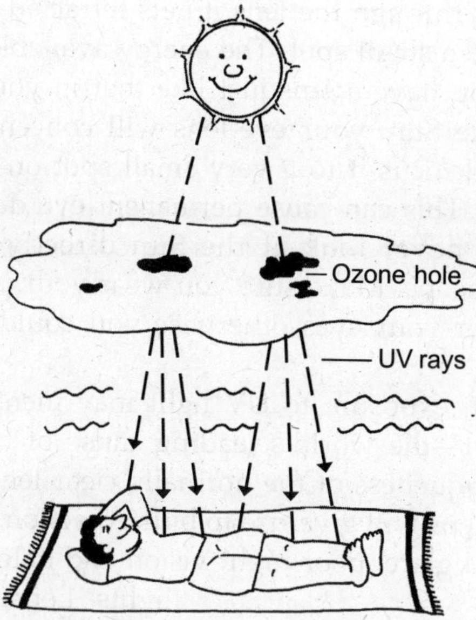

Fig. 6.6: When the ozone layer is damaged, UV rays from the Sun reach the Earth harming both health and environment

dangerous. Stay out of the Sun between 12 noon and 3 p.m. as much as possible. Hats and tightly woven clothing protect the shielded body areas from exposure. A broad-brimmed hat provides protection to the face and neck. Check labels for sunglasses that block most of UVA and UVB light (Fig. 6.7).

To avoid excessive exposure to ultraviolet light apply Sunscreen lotions and creams with skin protection factor 15-30, which are labelled as effective in screening UVA and UVB, provide good protection when applied as directed. Apply more often to nose, feet, earlobes, and bald spots.

Eye Problems

As a child have you ever tried to take a magnifying glass out into the Sun and burn leaves? When the focused sunlight enters through the lens it gets refracted and gets concentrated to a small spot. The energy available is truly remarkable. You have a lens just like that in your eye. If you look at the Sun, your eye-lens will concentrate the Sun's light and focus it to a very small spot on the back of your retina. This can cause permanent eye damage or blindness. So never look at the Sun directly or even through a telescope. Make sure you wear your protective sunglasses over your eyes otherwise you could damage your eyesight.

Long-term exposure to UV light may increase your risk of cataract—the world's leading cause of blindness. Cataract is opaqueness of the normally clear lens of your eye. Clouded lens will give rise to blurred vision, sensitivity to light and glare, poor night vision and halos around lights. Lenses darken due to long-term exposure to UV light and the darkening in turn causes the lens to absorb more UV light. Normally, the lens filters out damaging UV light from the Sunlight and allows other light to pass through to your retina. But while protecting your retina, the lens is slowly self-destructing.

Fig. 6.7: Sun protection

UV light may encourage formation of free radicals. Free radicals are toxic molecules that injure cells in your lens. And since the lens never sheds old cells throughout life, damaged cells accumulate and the lens becomes less transparent. Although some degree of cataract formation is normal, as you grow older it may be the overexposure to UV radiation from Sunlight that is now an accepted cause of cataracts.

Scientists have found that people who are exposed to high levels of Sunlight are up to four times more likely to develop cataracts. It's thought that the ultraviolet component of solar radiation may speed up the clouding of the lens.

Heat Blues

Sun can turn into a foe for people who work outside for extended periods of time. The Sun can be disturbing to a lot of people who work in the open. Farmers working on the fields, fishermen at sea, all need to make provisions for the Sun's effects in relation to their work. Engineers especially have to take into account the Sun's effects on waves and electrical devices while working on projects. Athletes also have to take into consideration the effects of the atmospheric conditions while training. Differences in the Sun's heat can cause major problems for them. Heat disorders can even affect the Sunbathers (Fig. 6.8).

Factors like temperature, humidity, air movement, and in addition a person's weight, medical condition and acclimatisation to heat also affect the amount of stress people receive. High temperatures increase the blood circulation in the skin and thereby increase the skin temperature causing the body to give off excess heat through the skin. But when the muscles are being worked, less blood is available to flow to the skin and release heat. And this causes the body to overheat. The body cannot get rid of the extra heat so it stores it. The body temperature then rises and the heart beats faster. The person becomes sick, irritable, fatigued and may faint due to heat stress.

Other things may also happen. Sun may reactivate the herpes virus in people who are affected by herpes. An increase in temperature due to Sun may cause insomnia and sleeping problems. Sun may cause cardiac troubles in those who suffer from heart problems. Sun may provoke vasodilatation causing venous problems in people suffering from varicose veins. Excessive heat from the Sun can cause tiredness and weakness.

Heat Stroke

Heat stroke is one of the most common yet deadly effects of the Sun on the body. A heat stroke results because of a serious failure of the body's heat regulation mechanisms due to excessive exposure to intense heat. This makes sweating stop. When sweating stops it makes the body unable to get rid of extra heat. People who have a case of heat stroke often have a body temperature of 41.1 degrees Celsius and sometimes higher. They are often

Fig. 6.8: Sunbathing

mentally confused; they sometimes faint, or go into a coma. Their skin becomes hot to the touch and very dry.

It is true that different people react in different ways to heat. But it is mainly the elderly, infants, athletes, construction workers, miners etc., who are more susceptible to heat stroke than others. The susceptibility also depends on the differences in age, weight, physical fitness, metabolism, extent of acclimatisation, decreased fluid intake, consumption of alcohol and/or drugs before work in the Sun and various other medical problems.

Heat Exhaustion

Many a times when you feel the heat and you sweat a lot, you feel the exhaustion or get that tired feeling and feel giddy. Other signs of heat exhaustion include nausea, headaches, moist clammy skin, and pale or flushed skin. Too much fluid loss through sweat can cause this condition. Heat exhaustion is caused by a loss of fluid through sweating. Severe heat exhaustion can result in unconsciousness and the person would require medical treatment.

Heat Cramps

When you work continuously in heat or in the Sun your body may lose salt through sweating and you may suffer from heat cramps. These are painful spasms of the muscles. You can treat heat cramps by drinking liquids in moderate amounts.

Heat Syncope (Fainting)

Very hot environment or working in hot Sun can cause fainting or heat syncope. This is common in people who have newly shifted to a hot climate or have come back from a vacation etc. Lay down the person who feels like fainting. Movement will prevent people from fainting.

Heat Rash

Hot and humid environment can cause heat rash. It causes redness and may be painful. Avoid being in hot and humid environments for long to avoid heat rash.

Fluid Loss

Heat causes dehydration and is so very common. Very few of us go through life without being dehydrated at some time or another. Dehydration drains you of the water and fluid content, and the electrolytes—the essential body salts needed for normal functioning. It needs timely treatment. Whenever there is a 5 per cent fluid loss or merely mild dehydration, the person can be easily re-hydrated. About 10 per cent of fluid loss from the body is considered to be a moderate dehydration. But when about 15 per cent body fluid is lost, the person may be severely dehydrated and may even need hospitalisation and treatment with intravenous fluids.

7

SUN WILL SHINE UPON US

Dating as far back as the Sumerians, all the way up to the present, the Sun has had an impact on people and their lives. The Sun's influence has not only led to many scientific discoveries, it also has affected the Arts, including music, painting, and dance. It has played an important part in our development as a modern society and continues to influence us today and will do so into the new millennium. It is because of the Sun that life exists on Earth.

Today, we see a vast advancement in the technological development around the world. The discoveries and inventions that took several years in the past can be brought about within a short span of time. A technologically advanced world has even helped to raise standards of living. Globalisation is in due to the ever-growing use of the Internet. These technological developments and the transition to a culture that is more aware of the need to safeguard the environment has helped create a world powered by the Sun's energy. And yet the problems of ever-increasing population with ever-increasing consumption of energy resources driven by the diffusion of life-styles in the industrialised and modern society still pervade around the world.

Solar energy infrastructure, whether installed in remote rural areas in a developing country or integrated in existing conventional infrastructure in a city in the developed world, needs to be better known and accepted. In order to spread the use of solar energy to the advanced world, we need to involve solar scientists, engineers, environmental researchers, financial experts and architects. All in all we need to nurture a new generation of solar-energy believers. A new culture of energy efficiency can lead to a more concerned, socially responsible use of all natural resources that will help promote lifestyles and prosperity, which then will help meet all the challenges of the next century.

Appendix

Solar Eclipses (as listed by BBC)

When	*Where*	*Type*
4 Dec. 2002	Southern Africa, Antarctica, South Indonesia, South Australia	Total
5 May 2003	Europe, Asia, North America, Northern Scotland, Iceland, and Greenland	Annular
23 Nov. 2003	Australia, New Zealand, Antarctica, Southern South America	Total
19 April 2004	Antartica, Southern Africa	Partial
14 Oct. 2004	North-east Asia, Hawaii, Alaska	Partial
8 April 2005	New Zealand, North and South America, South Pacific Panama, Columbia, Venezuela	Hybrid
3 Oct. 2005	Europe, Africa, Southern Asia, Portugal, Spain, Libya, Sudan, Kenya	Annular
29 Mar. 2006	Africa, Europe, Western Asia, Central Africa, Turkey and Russia	Total
22 Sep. 2006	South America, Western Africa, Antartica, Guiana, Suriname, F. Guiana, Southern Atlantic	Annular
19 Mar. 2007	Asia, Alaska	Partial
11 Oct. 2007	South America, Antarctica	Partial

7 Feb. 2008	Antarctica, Eastern Australia, New Zealand	Annular
1 Aug. 2008	North-eastern North America, Europe, Asia, Northern Canada, Greenland, Siberia, Mongolia, China	Total
26 Jan. 2009	Southern Africa, Antarctica, Australia, South-eastern Asia, Southern Indian Ocean, Sumatra, Borneo	Annular
22 July 2009	Eastern Asia, Pacific Ocean, Hawaii, India, Nepal, China	Total
15 Jan. 2010	Asia, Central Africa, India, Myanmar, China	Annular
11 July 2010	Southern South America, Southern Pacific, Easter Island, Chile, Argentina	Total

Printed at : First Impression, Noida